Hermes and the Telescope

HISTORY AND PHILOSOPHY OF SCIENCE

Heresy, Crossroads, and Intersections

Paolo Palmieri
General Editor

Vol. 2

This book is a volume in a Peter Lang monograph series.
Every volume is peer reviewed and meets
the highest quality standards for content and production.

PETER LANG
New York • Bern • Frankfurt • Berlin
Brussels • Vienna • Oxford • Warsaw

Paolo Palmieri

Hermes and the Telescope

In the Crucible of Galileo's Life-World

PETER LANG
New York • Bern • Frankfurt • Berlin
Brussels • Vienna • Oxford • Warsaw

Library of Congress Cataloging-in-Publication Data

Names: Palmieri, Paolo.
Title: Hermes and the telescope: in the crucible of Galileo's life-world / Paolo Palmieri.
Description: New York: Peter Lang, 2016.
Series: History and philosophy of science: heresy, crossroads, and intersections; vol. 2
ISSN 2376-6336 (print) | ISSN 2376-6344 (online)
Includes bibliographical references and index.
Identifiers: LCCN 2016014986 | ISBN 978-1-4331-3140-0 (hardcover: alk. paper)
ISBN 978-1-4539-1632-2 (e-book)
Subjects: LCSH: Science—Methodology. | Formal discipline. | Galilei, Galileo, 1564–1642. | Science—Experiments—History—17th century.
Classification: LCC Q175 .P327 2016 | DDC 501—dc23
LC record available at https://lccn.loc.gov/2016014986

Bibliographic information published by **Die Deutsche Nationalbibliothek.**
Die Deutsche Nationalbibliothek lists this publication in the "Deutsche Nationalbibliografie"; detailed bibliographic data are available on the Internet at http://dnb.d-nb.de/.

The fighter
In the afterglow of the numinous encounter
With the psychedelic visionary, Terence McKenna,
A hell of a fighter!
The hermetic light shines through the loved child,
Borne by the wind in his belly,
Reborn in the flesh from the ashes of books,
Creature of time,
The iris of soul,
Cracker of the eternal code, incandescent whiteness.

Paolo Palmieri—From: *Catheterisms* (unpublished).

The termite
Good grief, if you met a termite wandering across
The floor of the jungle and interviewed him on his life's work,
And he announced that it was the discovery of certain truths,
You would be fairly condescending in how you related to that.
Well, but do you believe that you are greatly different in your cosmic positioning than that termite?
You know what, monkeys are better at this than insects?
I don't think so.

Terence McKenna—From: Weekend Workshop, Esalen: 27–29 August, 1993. YouTube Video.

CONTENTS

ACKNOWLEDGMENTS

The journey that brought me to the project of this book started long ago. Its ideal beginning was an act of sincere hospitality by which I was accepted as a graduate student in the *Science and Technology Studies* department at University College, London. There I met Andrew Gregory and Hasok Chang, who became my doctoral supervisors. They were instrumental in helping me develop ideas in a spirit of total openness and liberal inquiry that paved the way for my professional pursuits. I thank them from the heart. It behooves me to remember that it was thanks to an extraordinary man, Colin Norris, that I was able to perfect (within limits) my knowledge of English to a point that I could express myself fluently in a second language. He taught me more than the passion for the idiomatic expression, or the perfect collocation of words, or the nuances of meaning and phraseology. He taught me a style of thought. My undergraduate years in Modena, Milan, and Bologna were life-changing. The foremost historian Paolo Prodi at the University of Bologna warned me that he needed to free me of the straightjacket of the history of science. It was a healthy warning that I have never forgotten. My friends William Shea and David Wootton have been wonderful mentors, from whom I learned how to temper my excessive polemical fervor. Stefano Gattei inspired me to pursue lines of research in the margins of seventeenth-century culture that proved to be most important. Over the course of many years at Pittsburgh I was fortunate to share interesting conversations on philosophy, mathematics, cuisine, frivolities, and science with a motley crew of chivalry figures. I salute the paladins of the long seventeenth century, *Oh gran bontà de' cavallieri antiqui!*, Ted McGuire, Ken Manders, Peter Machamer, and the apostate John Norton. Sprightly images of intellectual freshness crowd in my mind. Among them, Tiziana Bascelli, Bryan Roberts, William Lebig, Emi Iwatani, Elay Shech, Eric Hatleback, Michal de Medonsa, Pat Corvini, Marina Baldissera Pacchetti, George Borg, Giovanni Valente, Nora Boyd, Haixin Dang, Siska de Baerdemaeker, Michael Miller, Rebecca Thomas, Zina Ward, David Pence, Jacob Neal, Mahi Hardalupas, David Colaco, Trey Boone, Jason Rampelt and many others whose names have faded in my memory. Above all I must be grateful to the countless undergraduates at Pittsburgh who have been exquisite partners in sharing the most genuine interest in my research and teaching. I am especially indebted, in a mysterious way, to Aaron Novick whose brilliant mind and lovely friendship conspired with me over the last few years. Vivian Appler has been a wonderful companion in exploring cross-disciplinary pathways of inquiry that challenged my imagination. While writing the manuscript I met the exilic poetess, Edith Doron, who was moved by my teaching emancipatory acts. Her diamond-sharp intellect and angelic grace illuminated my groping towards individuation, revivifying the anarchist that slumbered in my spirits. But this book is the confluence of many secret undercurrents and turbulent rivers, whose sources are in a happy childhood and family, wonderful parents, Anna and Giorgio, industrious grandparents, and the wonders of an inclusive, freely accessible, humanistic educational system that Italy may justly be proud of. Destiny or providence steered my life-course through the alchemical works of the feminine, incarnated in my wife, Paola, urging me to think otherwise, in deeds of liberty and in freedom of thought, sentiment, and will. Last but not least I warmly thank my friend Frits Jonker for the cover artwork.

PREFACE

The book invites readers to encounter Galileo in his life-world, which also means encountering myself within my own life-world. I call *life-world* the familiar surroundings or dwelling-place for human beings that constitute their originary being at home, though the life-world should not be conceived as a reality existing independently of them. Correlatively, human beings cannot be portrayed separately from their being-at-home in a life-world, which is constitutive of human at-homeness. However, this primordial being at home is mediated by living presences encrusted within memorial traditions that define identities, objects, and thresholds. The rational fabric of Galileo's life-world, ostentatiously certified by his sibylline motto, a yearning for *sensate esperienze e certe dimostrazioni*, is interwoven throughout with motifs of hermeticism. These motifs form a unique style of being-in-the-world, always poised between gloom and laughter. A jester survives in the comedy of heresy and libertinism, and yet a thread is traceable delineating a quest for the divine. The book responds to my Herculean labors to reenact Galileo's experiments—a work that has occupied me, emboldening and frustrating, for two decades. The lesson I have drawn from that engagement with materials, reinvented apparatus, and reticent texts is that the role of experiential learning within Galileo's life-world is intractable from historiographic perspectives that focus exclusively on textual exegesis. The latter must include an understanding of the involvements, his and ours, that are constitutive of the significance of the totality of his and our claims to the veracity of lived-through experience. Reenactment suggests that a tradition of textual memorials becomes intelligible only in a confrontation with life that embraces wholeness, which means transforming oneself continually in order to be at home in one's transforming life-world. This conclusion is motivated by the stunning results of reenactment.[1] Texts do not encode the Mercurial phenomena of lived-through experience. The veracity of lived-through experience transcends the descriptive power of language. The face-to-face with objects on the scene of experiment poses the problem of severance in the knowing process, the hiatus between the verbally articulable and the unspeakable, namely, the experiential. The exegete who withdraws from the scene is marooned on an island of deaf mutes. He lives in a state of destitution of the imagination that emasculates the heroics of hermeneutics. This is why, I think, Galileo's experiments have been so often misunderstood, or their historical reality denied. This book offers a critical reflection on my attempt to restore Galileo's life-world within my own life-world, or vice versa, on the mystery of my feeling at home in Galileo's.

Most historians, sociologists, and philosophers of science, including my past self, have portrayed Galileo as a contributor to the scientific revolution.[2] We have tended to privilege micro-history and a conceptual framework that speaks to analytic philosophy, or the sociology of science, regardless of the differences concerning the minutiae of the specific objects that were put under scrutiny. The unintended consequence of this wide-spread methodology has been the neglect of esoteric aspects of Galileo's life-world that challenge the positivist schema of the emergence of modern science. However, Nicholas Jardine has warned that "it is misleading to credit Galileo with promotion of a modern mathematical-cum-experimental science".[3] I agree with his assessment, in the sense that friendly agreement gives me the opportunity to think with the freedom of suspending the prejudices of received historiography and of questioning my prejudices, by way of playing with them for the pleasure

1 I take the liberty of referring the reader to Palmieri 2008, 2011, for a more thorough discussion of the reenactment of experiments in a philosophical perspective.

2 David Wootton (2015) has given us a magisterial account of the scientific revolution, charting the many intellectual and cultural territories that underwent dramatic transformations in early modern Europe, and which later on culminated in the advent of modern science.

3 Jardine 1991, pp. 101–102.

and the sake of it, as if there were a deeper truth to be glimpsed when thought plays with itself. In this playful spirit, I read his admonition as an implicit invitation to reappraise the analytic framework that has prevailed in the field. Thus I have taken an integrative methodological stance while writing the book. I label it *hermeneutic encounter*. I summarize it briefly, to myself and to the benevolent reader, as follows.

I look at historical documents from a perspective that is hospitable to the ethnicity of reason (on which, I will say more below). The analytic framework aims at formulating logical arguments in support of historiographic or philosophical claims. In first approximation the hermeneutic encounter aims at experiencing the outgrowth of complex historical objects from their roots in the life-world. Within the life-world dereistic energies such as paradox, incoherence, or estrangement, nurture the soil in which creativity germinates. The notion of *dereistic energies* refers to modes of being-in-the-world that do not privilege one particular reality, or a particular formal order such as the polarization of knowing subject and object of knowledge. Dereistic energies disinhibit creative intercourse within the life-world. The hermeneutic of encounter empowers the reader to experience these active forces. It rehabilitates the epistemic value of paradox, incoherence, or estrangement, which only appear to be marginal, if not degenerative phenomena from the traditional viewpoint. I work by placing sundry historical sources on a larger historical canvass. First I contemplate them as uncooked materials in an alchemist's crucible. Then I fire the furnace to vivify the mixture in my imagination. This iterative process of dissolution and coagulation of ideas (*solve et coagula*, the secret of the spagyrist) evokes formative realities as diverse as, say, astrological lore, or devotional pilgrimage, or the grotesqueries of Olympian deities on the stages of the *commedia dell' arte*, or religious heresy. It depends on the ingredients in the crucible. Finally, I put the transformed materials to the test of encounter, letting their souls blurt out strange words in their own dialects. Thus I enter into a dialogue with them from within the horizon of my own life-world. The alchemy of the imagination reactivates the dereistic energies originally present in Galileo's life-world, allowing me to synergize with his creative processes.[4] Active imagination has the power to transcend the culturally

4 Some might balk at the idea that imagination takes center stage instead of deductive reasoning (without of course negating the latter), but in Galileo's life-world imaginative thought-experiments are often more important than deduction and, even more astoundingly, they have proved to be irreducible to formal patterns of reasoning a priori, or to induction (Palmieri 2003).

sanctioned censorship of consciousness and corporeality, and thus attain a more genuine openness for being-in-the-world, towards and beyond the finite world horizon. This power of the imagination that resides beyond individual consciousness, with its marked pantheistic inspiration, was well known to the hermetic-spiritual alchemists of the seventeenth century, but it has since been obscured by Christian rationalism, the notion of truth as *adequatio intellectus et rei*, that quickly allied itself to early modern science, because it threatens Christian eschatology, i.e., the doctrine of the individual human soul that will face God's final judgment.[5]

In the chapters I collate the products of this alchemy of the imagination. Their sequence is not intended to form an overarching deductive structure though fragments of aporetic argumentation will inevitably be intermingled in the text. It is the alchemy of the imagination that sustains the hermeneutic encounter. The chapters are *canovacci* (like the plots of the *commedia dell' arte*), each featuring a principal character, or archetype, lending the chapter its evocative title. These archetypal figures were coaxed out of the crucible of Galileo's life-world. They are imaginative transmutations of dereistic energies that I could resonate with. But Galileo's life-world is not a territory open for exploration in a straightforward sense, for, as I already intimated, it is not an object that exists separately from my own life-world. Rather than being an adventurous explorer, I therefore prefer to play the host in a host-guest relationship that catapults both figures into the drama and revelation of hospitality.

There are features in the hermeneutic encounter which break with certain expectations. I reach no definitive conclusions that could be demonstrated in a

---

5 It was splendidly portrayed by Michael Sendivogius, who thought that it has the same creative power as God's imagination, beyond the confines of the human body, though limited to the realm of mind. "Anima autem qua homo a caeteris animalibus differt, illa operatur in corpore, sed maiorem operationem habet extra corpus, quoniam absolute extra corpus dominatur, et his differt ab animalibus, quae tantum mentem et non animam deitatis habent. Sic quoque Deus et conditor rerum, dominus et Deus noster operatur in hunc mundum ea quae ad mundi necessaria pertinent, et in his includitur mundo, unde Deus ubique esse credendum. Excluditur autem illa sua immensa sapientia, extra corpus mundi qua operator extra illum, et alia multo altiora imaginatur quam corpus mundi concipere potest, et ista sunt extra naturam, solius Dei secreta: sicut est exemplum in anima, quae extra corpus multa profundissima imaginatur et hisce assimilatur Deo, quibus extra suum mundum operatur extra naturam, licet et ista sunt adhuc candela accensa erga meridianam lucem, quoniam anima imaginatur sed non exequitur nisi in mentem, Deus vero omnia eodem momento quo imaginatur efficit" (Sendivogius 1628, pp. 143–144).

stringent form of argument such as a demonstrative syllogism. I find reasoning that seeks apodictic closure at all cost a degenerative pathology of mind that has affected my way of thinking, through both education and cultural conditioning, but which is not incurable. The received justification for seeking apodictic closure is supposed to be an ethical norm, the intellectual honesty that the investigator proclaims as unrenounceable commitment to eternal truth. But all to often, it seems to me, this justification is a mask hiding an inflated egotism that we have inherited with vengeance from the male-dominance hierarchies in our primate lineage. Behind it is the principle of pleasure. It is the pleasure of witnessing, or engaging in the violent diatribe in which a designated defendant is finally slaughtered in the public arena before a conniving audience.[6] Hence a measure of relief, if not an impossible redemption, must be sought somehow, for human experience needs to be restored to the balance of health, which entails the recognition of our infinite obligation to all forms of existence. Throughout modern European history the violence of the controversialist, religious, political, intellectual, has all too often translated into the brutality of the fascist interrogator.[7] Argument is a violence that coerces the litigants into an implosive, self-destructive state of mind, into a violation of each other's at-homeness in their life-world. It denigrates whatever is perceived as resisting its belligerent logic.[8] The hermeneutic encounter that

---

6 But, one might certainly ask, isn't any positive statement, in the form subject-predicate, an apodictic affirmation claiming a truth, and hence exercising the very violence you are deprecating, if only against an unspecified interlocutor, perhaps the anonymous reader, and consequently are you not deluding yourself that the violence of argument can be at least subdued, if not eliminated? Possibly, I am at a loss, for I feel the dilemma. A mitigating answer might be as follows. I am not seeking closure when I make a positive statement, but I am looking for a response to the mysterious and frightening appeal from the abyss of the relation that being-in-the-world entails towards the 'there' in which I always find myself entrapped; therefore, I must welcome others seeking their own responses, potentially at variance with mine, without enforcing a common denominator. I am under the obligation to welcome the infinite difference stemming from the abyss, and can only hope to be welcomed as a guest by others.

7 I have been the victim of this violence multiple times, and I have been its perpetrator, a torture which is particularly despicable in the form of intellectual mobbing, as when the collective rage of the majority seeks gratification in 'winning the argument', in slaying the deviant individual who does not 'listen'. We are all the infamous teacher of Ionesco's *Lesson*.

8 There is a farcical element in this tragicomedy of philosophical violence. It is the riot of questions and answers that I have frequently witnessed at philosophy conferences, the public lynching of the unfortunate presenters, a veritable *panem et circenses* that offers

I propose does not reconstruct historical facts by way of violence, at least not consciously, or in order to demarcate orthodox styles of inquiry from deviant ones. Rather, it seeks to decipher the spectrum of involvements diffracted by the luminous presence of the guest. It entangles the object-guest with the nakedness of the host, revealing the queerness of their relationship. The renunciation of closure opens a door to the silent being of objects. For, as the poetess, Edith Doron, writes, "world, as that which is hidden in plain view, must be torn asunder, brought to the fore in order to free the being of that silent object which has 'agreed to keep quiet about us' from an imperial interpretive statement. If we are to understand encounter not as an act of comprehension in the sense of 'grasping the world' but instead as the *immersive experience of de-familiarization*, a dislodging of the concealed given ground of world, then the possibility of transformation is freed".[9] Thus, I hope the reader will be

---

the gladiators (especially the *damnati ad ludos linguaces* from the frontier of the analytical Empire) the chance to make a killing. The flamboyant Athanasius Kircher described the spectacle admirably. I will not translate the passage because I would spoil the fantastic laughter I promise the reader. Kircher 's Latin is hilarious. "Haud dubie erant, qui dum sermonem proximi non intelligerent, se illudi putantes, non verbis, sed verberibus pugnisque responderent; ex rixis enim ortis, dum pro bitumine lateres, laters pro caemento, scalas pro lintribus, instrumenta ferrea pro foeno et paleis, carnes pro caseos, caseum pro fructibus porrigerent, in indignationem versi, qua fustibus, qua utensilibus instrumentis egregie sibi capita pertunderent, ut proinde hac prodigiosa obtorta confusione, sacrilegos hominum, nefandosque ausus, DEUS iuste compescuerit; ut qui contra DEI voluntatem aedificium illud extruere conarentur, illi confusione linguae, a coepti operis molimine desistere cogerentur" (Kircher 1674, p. 125).

9 Doron 2008, pp. iii-iv. Emphasis is mine. I would like to illustrate what I mean by *experience of de-familiarization* with the example of a revelation that occurred to me. Long ago Antonio Favaro published a fragment of a curious document preserved in the *Archivio di Stato* in Modena, dated June 25th, 1633, in which Antonio Badelli, a contemporary reporter, communicated from Rome that Galileo had abjured (as he did on June 22nd), and that during the ceremony his book (the *Dialogo*) had been burned in his presence. "...Il Galileo fu abiurato mercordì mattina nel Convento della Minerva alla presenza di tutti i Cardinali della Congregazione, e gli abbruciorno in faccia il suo libro, dove tratta del moto della terra..." (Galilei 1890–1909, XV, p. 164). The ellipsis left by Favaro at both the beginning and end of the short passage tickled my curiosity. At that time I had no special interest in the Galileo trial as I tended to believe the positivist myth, the story of the great scientist victimized by the Roman Inquisition, whose case became the most flagrant episode of intolerance in the confrontation between science and religion. However, my instinct brought me to Modena, for I wanted to see the original and taste the pleasure of handling with my hands the piece of paper, almost four hundred years old, that carried the news of such a momentous event in European history. When the archivist fetched the dusty box containing the collection, I looked for the newssheet. I started reading. I was

willing to join the jam session and extemporize on the schematics of action and thought that are offered in the *canovacci*, as enjoyable opportunities for encounter, for then the magic of transformation will be realized. My imaginations remain open and enthymematic, at times conflicting with one another, like melodic cues, dissonances waiting for a reader's brilliant response, so as to depart anew in unforeseen directions. The archetypes occasionally recede, shadowy and ambivalent, and the flow of energy that returns the elements to *massa confusa* (primordial chaos) would require re-visioning the archetypes. I have striven to subdue my argumentative mindset, trying to resist the model of intentionality borrowed from the natural sciences that exacerbates the violence of the inquisitor. I have assembled the *canovacci* not so much from the perspective of the axiomatic ego, enjoining a foreclosure of mind, as from that of the exiled self, soaring in the nocturnal emptiness of encounter and willing to be lost in the immersive experience of de-familiarization.

The strength of the analytic method is that it posits particular historical objects as given at the beginning of inquiry, typically those studied by micro-history, such as Galileo's law of falling bodies, or the discovery of the moon's craters, to mention a couple of well-known examples. This givenness of the objects creates a starting-point for analytic reason to effect the dismantling of the objects into structural components that are taken to be explanatory (in some sense) of the formation of the whole. It subsequently reconstructs the putative logic that is presumed to be inherent in the objects as they have been handed down, be it the logical-mathematical argument that allowed Galileo to derive the law of falling bodies from first principles, or the social history of artisanship, labor, and embryonic capitalism, encrypted

---

flabbergasted, and violently de-familiarized from the world that was my immediate cultural surrounding. My familiar world began to give way. The Galileo news was indeed cited, but only after Badelli had painstakingly chronicled the details of what was clearly perceived to be the most dramatic story of the last few weeks in Rome. It was a frivolous diplomatic dispute involving French and Spanish ambassadors, the pride of the Barberini, both the Pope (Urban VIII) and his family entourage, a boy and his tutor, the diplomatic abilities of Cardinal Guido Bentivoglio (a signer of the condemnation of Galileo on June 22nd), and a series of misunderstandings about the etiquette of a dinner party and a horse-riding event. The impression I gathered from Badelli's newssheet was that the abjuration looked like a no brainier for anybody involved (except for Galileo, of course), a piece of ordinary administration of justice, inconsequential in the larger social and political context. The Galileo case was never a case, as I pondered while driving home, but had become one in the fabulous amplifications of history. As far as I know this de-familiarizing document has not been taken into account in the diatribes and chest-beating display over Galileo.

in the technology of polishing precious stones that was perfected by the Venetian artisans, and which inspired him to build his first telescope. Very often, however, the historical objects posited as worth investigating within the analytic framework are the cultural heritage of nineteenth-century science and positivism. The ideology of the positive sciences that colonialist Europe opposed to the magical thinking of 'uncivilized' societies had racist implications that have remained unexamined in the field of the history and philosophy of science. These prejudices have restricted the scope of inquiry. I am not suggesting that scholars, including myself, have been consciously racist, rather I am suggesting that nineteen-century racist prejudices have been molded into our research and educational institutions. For instance, most scientific journals reject submissions in other languages than (typically) English, and many scientific and educational institutions tolerate or passively encourage ignorance of languages, one of the most pernicious manifestations of institutional racism.[10]

The analytic process of inquiry in the history and philosophy of science does not seem to have been efficacious in questioning the racist myth of

10 The *Report of the Stephen Lawrence Inquiry*, sponsored by the British Government and published in the United Kingdom in 1999, in the aftermath of the racially motivated murder of Stephen Lawrence, should be remembered in this context. "If racism is to be eliminated from our society there must be a coordinated effort to prevent its growth. This need goes well beyond the Police Services. [...] Just as important, and perhaps more so, will be similar efforts needed from other agencies, particularly in the field of education. As we have indicated, the issue of education may not at first sight sit clearly within our terms of reference. Yet we cannot but conclude that to seek to address the well founded concerns of minority communities simply by addressing the racism current and visible in the Police Services without addressing the educational system would be futile. The evidence we heard and read forces us to the conclusion that our education system must face up to the problems, real and potential, which exist." (MacPherson of Cluny 1999, paragraph 46.34). There is no need to emphasize that science is a most important component of Western educational systems. Although the phrase 'institutional racism' is open to different interpretations, the report noted in this connection that "[u]nwitting racism can arise because of lack of understanding, ignorance or mistaken beliefs. It can arise from well intentioned but patronising words or actions. It can arise from unfamiliarity with the behaviour or cultural traditions of people or families from minority ethnic communities. It can arise from racist stereotyping of black people as potential criminals or troublemakers. Often this arises out of uncritical self-understanding born out of an inflexible police ethos of the "traditional" way of doing things. Furthermore such attitudes can thrive in a tightly knit community, so that there can be a collective failure to detect and to outlaw this breed of racism" (*ibid.*, paragraph 6.7). There is indeed an uncritical self-understanding of science, and a scientific ethos of the 'traditional' way of doing science and science education.

science as the supreme court of knowledge. In sharp contrast, hermeneutic encounter is willing to acknowledge the institutional racism that has become part and parcel with our intellectual practices, including science, and hence to start a process of reparation aimed at integrating itself, on a humble level field, with the ethnical backgrounds of reason. It invests no objects, cultures or languages with special rights at the beginning of research. It renounces the principle that an absolute dialectics transcending the hermeneutic process is to be discovered within historical objects assumed as worth investigating. Rather, it mobilizes a motley assortment of world-less materials and objects, in the sense that they initially appear in self-reflecting consciousness as withdrawn from the world which originally constituted their meanings. It endows them with equal rights to enter into a dialogue with us. The alchemy of the imagination pursues their possible transmutations in order to liberate them into the totality of significance for universal existence. This non-violent hermeneutics of liberation reawakens the historical determinations that marked off the possibilities of being in which life has been involved in order to lay those possibilities open once again for involvement.

While writing in English I am well aware of the potentially racist import of European languages. To alleviate this concern I offer the humbling experience of having been a committed learner (I reject the title of professor) in multicultural settings for more than ten years, in which I strove to become responsibly attentive to the idiomatic expressions of ethnicity. It is not so much that ethnic reason responds differently to the encounter with the world-less objects of inquiry obsessively imported into the classroom by the academic instructor, as that it is captivated in wondrous life-worlds which are far removed from that of the academic instructor. Thus, I have renounced the search for a common measure that might establish a ground for ethnic integration in the multiethnic classroom as this ambition presupposes that there is such thing as a neutrality of reason. Invoking neutrality is tantamount to rejecting the genuineness of ethnic diversity insofar as integration only extends the borderlines of tolerance, but what is reason engaging in when it *tolerates* otherness, if not in muted racism? Since I do not tolerate but indeed *value* the ethnicity of reason, I let it speak in its own mother tongue, welcoming its sacred strangeness, both in the classroom and while doing research.[11]

---

11 I cannot expand here on the *pedagogy of the sacred* that I have developed over the past few years to overcome the racism of tolerant reason institutionalized in Western educational systems. But I will detail the work of learning in ethnic environments in a future study devoted to the ethnicity of reason and the experience of the sacred as learning process.

I study most primary sources in their original languages (though I do not suggest that this is a unique feature of hermeneutic encounter), translating or summarizing them for propaedeutic accessibility. But I am conscious that the act of translating is always an act of betraying. Often I quote the originals as accurately as possible, as welcoming gesture of *intolerant* submission to the reader, within the limits of my modest linguistic competence. I have already given an example at the beginning of this section, reporting Galileo's celebrated motto, *sensate esperienze e certe dimostrazioni*, without attempting to translate the phrase. There is no word in English that adequately renders Galileo's *esperienza*, the most mysterious entry in his personal lexicon, and its entanglement with the *certe dimostrazioni* makes it even more perplexing.[12] Perhaps "lived-through experience and geometrical proof" might do the dirty trick, but let the reader decide after pondering the plots I weave in the *canovacci*.

Throughout the book I take strolls within Galileo's life-world. I follow trajectories that dynamically converge towards hermetic figurations but sometimes border on heresy and libertinism. I rescue from oblivion the significance of hermetic being-in-the-world as open to the possibilities of understanding that are projected unto the heretic and the libertine. I thus monumentalize hermetic being-in-the-world. I point the reader to its restored potentiality for offering hospitality to ethnic styles of reason. I have encountered this unstable triangulation between religious heresy, libertinism, and hermeticism while reworking the categories of Italian *hereticality* and *libertinism* discussed in ground-breaking studies by Delio Cantimori and Giorgio Spini.[13] Renaissance hermeticism has never been put under scrutiny as a historiographic category capable of subsuming and therefore transubstantiating religious heresy and libertinism. I think it is worth exploring this line of inquiry. It promises to destabilize the cultural conditioning that keeps at bay the objects that have long been quiet about us.

Thus a picture emerges of Galileo's life-world, a totality of involvements in which the transformative power of hermeticism might loosen the soul from her attachment to the dark clothing of matter, earning her immortality, while holding out to the hermetic human a therapeutics of the sacredness and veracity of experience. This experiential hermeticism, which need not have been enshrined in an oath of allegiance to a doctrinal corpus, let alone formally defined in a pseudo-religious credo, has escaped attention, I suspect, precisely because the objects and modalities of inquiry that were privileged by analytic

12 Schmitt 1969.

13 Cf. Cantimori 1992 (1st ed. 1939) and Spini 1983 (1st ed. 1950).

investigators had been inherited from nineteenth-century positivism.[14] The historic *National Edition* of Galileo's works (completed in 1909), which has been the basis of Galileo scholarship until the present, was a grandiose monument erected by positivist historiography in the aftermath of the struggle against the Pope's temporal power that had inebriated the Italian *Risorgimento* (the patterns of warfare, diplomacy, and insurgence that transformed weak regional states into a politically unified nation). It has shaped our image of Galileo for more than a century. Its editor, Antonio Favaro, conjured a mesmerizing fantasy out of dusty archives. The father of modern science was born, or perhaps still-born, a martyr of the experimental method's struggle for survival, whom Counter-Reformation Catholicism subjugated to the demands of theological orthodoxy. The phenomenon of hermeticism, however, largely exceeded, if not violated, both Catholic orthodoxy and early modern norms for the production of *scientia* (as codified, for instance, in the logical teachings—such as the theory of *regressus*—of the Paduan Aristotelians studied by Pietro Ragnisco and John H. Randall).[15] In my view, it is best understood as a manifestation of the multilingualism of the Italian Renaissance. The centrifugal fragmentation pressurized by regional vernaculars that had loosened the tongues of Dante, Petrarca, and Ariosto, emptied the consciousness of professorial academics, while the second nature of Latin humanism infected

14 A notable exception to this tendency is the work on Newton and hermeticism by James McGuire (cf., for instance, McGuire 1977). Commenting on the multifarious composition of the Hermetic Corpus, McGuire emphasized that "the basic structure of the *Corpus* (and its underlying metaphysics) is Neoplatonic". His striking conclusion was that "[i]n general, Hermeticism, however understood, has never at any period had an intellectual life of its own; it is invariably found in association with one of the dominant traditions in Western thought, usually Neoplatonism. It does not provide either a body or a mode of thought that is viable without, or ultimately distinguishable from, any of the variations of Neoplatonic thought" (McGuire 1977, p. 127). I find McGuire's suggestion that hermeticism needs a breeding-ground in other modes of thinking intriguing. It points us in the direction of the mundane plane of existence where I desire to locate the hermetic phenomenon that is hidden from plain view. Although in his paper McGuire appeared skeptical about the direct influence of the Corpus Hermeticum on Newton, in conversation with him over the past decade (as I was his colleague at Pittsburgh from 2003 until his retirement) I formed the impression that his attitude to hermeticism was more complex. He taught a popular course for undergraduates that he had designed from scratch, called *Magic, Medicine, and Science*, and in which he incorporated many hermetic motifs into a broad narrative that linked Newton to the various strands of the Neoplatonic tradition.

15 Palmieri 2007 (with ample discussion and bibliography on Pietro Ragnisco as a precursor of John H. Randall).

a plethora of hackneyed literati in the service of despots. No wonder that Galileo chose the vernacular as his preferred place of habitation.[16] From this linguistic perspective Galileo's hermetic visions, memorialized in his comedic vernacular and tinted with the colors of Ariosto's Neoplatonic poetry, can be seen as a quest for individuation, the liberation of a spiritual Mercurius (alluded to in the title of Galileo's *Sidereus nuncius*, or *Starry messenger*) from the darkness of the material body. This hermeticism in the flesh is borne out by Galileo's reveries of light and of the infinite rarefaction of matter, an Ariostean flight of the alone to the alone over and above the great divide between religious heresy and orthodoxy, between Giordano Bruno and his prosecutor, Cardinal Roberto Bellarmino. On Galileo's reveries his pupils collected unequivocal testimonials, as we shall see. In a nutshell, within Galileo's life-world, hermeticism was being-in-the-world towards achieving the sublimation of the human soul, the immortality of being God in the all-pervading light of the divine presence.

Finally, I must point out that there is also an autobiographical rationale for this book. Over the last decade I regularly taught an undergraduate class called "Galileo and the creation of modern science". My teaching practices are unconventional. They are informed by a philosophy of pedagogy inspired by Maria Montessori that emphasizes learning as a holistic process of growth of both the individuals including the teacher and the class as a whole. A few years ago I started a laboratory where, with the help of students, I have recreated landmark experiments in the history of science. We performed experiments in the laboratory (such as the inclined plane) and imaginatively reconstructed Galileo's ideas. In the classroom, I have the students practice the watercolor techniques that Galileo employed to paint washes of the moon from memories of the occult iridescences he observed with the telescope. I bring their attention to the alchemical process of mixing water and pigments, and to the emergence of patterns of light and dark (*chiaroscuro*) on the white paper while the pigmented liquid is transmuted into air. I emphasize transmutation as opposed to evaporation since the latter concept cannot be applied to the seventeenth-century experience of the four elements. Then I invite the students to discuss the aesthetics of *chiaroscuro*, and to reflect on how aesthetic ideas become constitutive of the human mode of vision. Thus the idea of the book is rooted in the social learning experience with my

16 Hall (2013) offers a survey of Galileo's bed-time reading habits, but for philosophically robust monographs on the literary surroundings of Galileo's vernacular habitation, there is still nothing better to consult than Vaccalluzzo 1896 and Wlassics 1974.

students over the years. I wrote it bearing in mind their surprising reactions to activities and discussions. Social learning has been instrumental in steering my attention towards uncanny presences in Galileo's life-world. Questions emerged in the classroom concerning the self, the scene of experiment, the role of the imagination, humanism, the relation word/thing, and the technologies for the manipulation of the psyche invented by Renaissance magicians. The book does not restate current knowledge though it dialogically engages relevant scholarship. It is not tied to the pedagogy that informs my teaching. I have taken into account materials at the crossroads of early modern European history, literature and science, the philosophy of science, the history of art, and theater. But it remains unaccomplished insofar as I have not renounced the ambition of surrendering my prejudices.

· 1 ·

MYTH

Over the last few years my approach to the history and philosophy of science developed from an analytic style of thought, which emphasized the logical-mathematical structures of scientific argumentation, through an appreciation of the pragmatics of science, to the realization that both the subject and the object of investigation, my self and the other, the stranger, in all its protean appearances, are productively engaged in a dialogue that is constitutive of their being-in-the-world across history. It is a process of reciprocal individuation that cannot be scrutinized solely from the perspective of analytic reason. I have come to think that reason, or soul, is given in gradations, extending from human beings all the way through the animate and inanimate kingdoms and embracing the life of the cosmos. There is nothing original in this paradoxical feelings. I share the captivation with paradox with many strangers, such as Galileo, Giordano Bruno, and the pantheists inspired by him.

In previous work I have investigated scientific processes that lay claim to universal validity by appealing to rational norms that now appear to me to be historically variable. These norms evolve but do not converge towards a cohesive ideal. They remain fragmentary and, as I see it, intertwined with political and economic struggle, as well as with racial and religious conflict and mundane careerism. The transformation of the normative structures of scientific processes, the rationality of science, is in itself not graspable from the rationalistic-normative viewpoint that privileges the language of mathematics. This abstract rationality is a surface phenomenon that nowadays often manifests itself in the form of intellectual pretend-play, in ludicrous academic rituals such as tenure and promotion, where the total amount of grant-money one is worth (which is proportional to the overhead-money received by the university hosting the grantee) decides one's future and standing in the hierarchies of scientific communities. The deepest currents of scientific organization and production are vortical phenomena that remain scientifically intractable, and scientific research itself does not lead to the emergence of superior knowledge. The scientist is not a superior human type.[1] For all its secular posturing Western science has remained obsequious to Christianity in all of its fundamental epistemic commitments. It has explored the technological realization of the metaphor of the material universe more aggressively than other cultures, but its obsession with the uniqueness of truth as the adequation of mind and world (*veritas est adequatio intellectus et rei*), which it has inherited from late-medieval scholasticism and ultimately from Judeo-Christian monotheism, has contributed to genocidal colonialism, economic exploitation, and ecological catastrophe on a cosmic scale.

For two decades I have endeavored, more and more consciously, to harmonize hermeneutics and empiricism, i.e., textual exegesis and reenactment of experiments. Above all, I have been interested in the existential and humanistic dimensions of learned practices emerging in particular historical settings, especially the early seventeenth century, unto which, for reasons that I still do not understand, I inexorably project my possibilities of understanding, my fabulous being at home, my everydayness and aesthetics. In this chapter, I will

---

1 "It is generally admitted that with woman the powers of intuition, of rapid perception, and perhaps of imitation, are more strongly marked than in man; but some, at least, of these faculties are characteristic of the lower races, and therefore of a past and lower state of civilization. The chief distinction in the intellectual powers of the two sexes is shewn by man attaining to a higher eminence, in whatever he takes up, than woman can attain—whether requiring deep thought, reason, or imagination, or merely the use of the senses and hands." (Darwin 1871, II, pp. 326–327). These aberrant views were put forward as 'science' by Charles Darwin more than a century ago. It would be naïve to think that contemporary science has been freed of these aberrations, for the fact is that Darwin's evolutionary theories cannot be dissociated from the racist worldview that made them appear credible and rational to educated Europeans, and which is presupposed by the concept of the superiority of scientific knowledge, which is rarely criticized in the public understanding of science and permeates educational institutions.

discuss the motivations that pushed me to scrutinize my own philosophical trajectory. I will raise questions concerning the analytic style of thought that has prevailed in Galileo studies. Like in an overture motives will be heard that I will evolve in subsequent chapters.

The Myth

The myth of origins that has been handed down to us by the high priests of nineteenth-century positivism would have it that Galileo was the founder of modern mathematical-experimental science, and that his revolutionary achievement consisted chiefly in dismantling the constructs of the late medieval cosmos, that is, in eradicating its roots in the erroneous doctrines of Aristotle and of his late scholastic followers. Undoubtedly, there are elements of historical truth in this as in all myths. I am a naïve believer in the euhemeristic interpretation of mythological fables though I like to think that historical events are also as wondrous as fables. It is the intertwining of sundry materials, history and fable, that fascinates me. But the Galileo myth conceals deeper layers of meaning that are buried in our collective unconscious, and which can be brought to consciousness, for example, by examining the linguistic sediments deposited in Western literary traditions. Over the last hundred years or so, scholars have tended to perpetuate the image of Galileo in his heroic quest for scientific knowledge. They shared a common epistemic goal regardless of their social or intellectual orientation towards history. This image has featured the hero in different postures, as the trail-blazer of modern experimental method and the creator of mathematical physics, as the founder of telescopic astronomy who 'abolished' the Graeco-Christian heavens (as Brecht put it in his portentous play), but also as the uncanny appropriator of late-medieval philosophies of science rooted in the logical theories of Aristotle that had been elaborated at Padua in the sixteenth century. Thus the prevailing image of Galileo in twentieth-century academic culture has been that of the natural philosopher, or prototypical scientist, intent on pursuing rational rules of inquiry, by fair means or foul, in order to establish objective truths about the material universe.

I renounce the positivist legacy though I do not negate its validity as myth of origins, for it has become part and parcel with our cultural consciousness. I do not put myself on a pedestal. I still enjoy the myth. Rather, I implicate my inquiry within a process of appropriation of the deeper meanings that can be recovered from that legacy by reconfiguring the mythologems. I understand Galileo's hermetic labors as a search for immortality within a life-world which was permeable to conflicting cultural currents, and in which conflict was dramatic, violent, and devastating though often it corroded the life of consciousness silently. From this irenic perspective, inclusive rather than exclusive, Galileo's natural-philosophy can be seen as an instrument of personal healing, i.e., not so much as a rigid doctrinal corpus but as a continual edification of his own individuality in a religious perspective. The labors took the form of an inquiry into nature but the hidden complex behind them was a longing for self-knowledge, or, to borrow an alchemical metaphor, for the liberation of a spiritual Mercurius, Galileo's own self, from the darkness of the material body towards immortality. My strategy is to dissolve Galilean mythologems and excavate the deeper strata of European culture from which the myth emerged. I spotlight the intersections of Galileo's existential pathway with early modern traditions that have been rejected or marginalized, or disqualified as esotericism, by the rise of the natural sciences in the nineteenth century. I portray a Galileo whose processes of self-edification point us in unsuspected directions. A picture will emerge of the shifting borderlines between hermetic work and the religious dilemmas of early modern Europe. The project is not entirely new.

Galileo's 'Strange' Opinions

About a century ago, Antonio Favaro published a number of 'strange' opinions and theories that were attributed to Galileo by his closest friends and disciples, but which had, so to speak, fallen by the wayside of history.[2] Favaro was a staunch positivist opposing the Catholic church in the aftermath of the political unification of the Italian state in 1861. His positivist faith prevented him from making sense of the allusions to Galileo's hermeticism that he had unearthed. In the *Domus Galilaeana* at Pisa—an institution founded in 1941 by the philosopher Giovanni Gentile under the aegis of the Italian fascist government, with the goal of promoting historical research on both Galileo and the "scientific movement originating with him"[3]—revealing documents

---

2 Favaro 1907, 1949.

3 "Per iniziativa della Società Italiana per il Progresso delle Scienze veniva istituito nel 1938 un comitato presieduto da Giovanni Gentile, con il compito di elaborare un progetto per la costituzione di un istituto in onore di Galileo Galilei. I risultati dei lavori del comitato, illustrati da Gentile il 15 Ottobre 1939, nell'Aula Magna dell'Università

are preserved from the archive of the Favaro family. Among them there is a fascinating collection of posters that were printed in 1897 on the occasion of a social and political demonstration in Pisa against the Catholic Church. They epitomize Favaro's philosophical and political outlook.

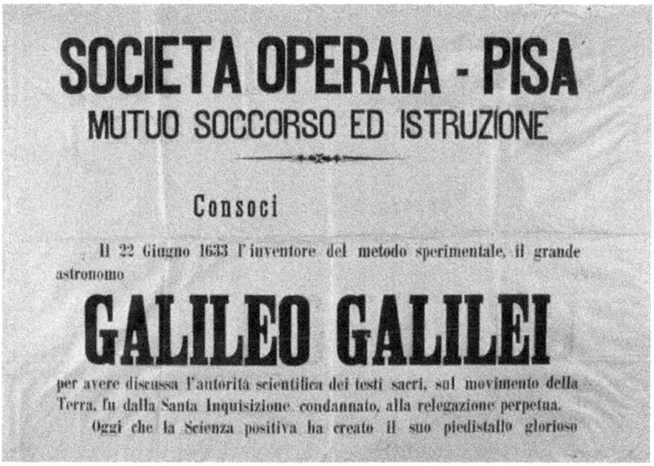

Fig. 1.1. The *Pisan Workers Benefit Society and Education* poster published in 1897 in honor of Galileo. (The poster was probably cut as the last sentence appears to be truncated). From: Fondo Antonio Favaro. Fondazione Domus Galilaeana, Pisa.[4]

The picture above (fig. 1.1) shows one of the posters. It was issued by the "Pisan Workers Benefit and Education Society" (*Società Operaia Mutuo Soccorso e Istruzione*). It addresses Galileo as the inventor of the experimental method (a turn of phrase for which Favaro himself had great fondness), as the great astronomer condemned by the Roman Inquisition for questioning the scientific authority of Scriptures on matters concerning the motion of the earth. The last sentence reads: *Today that positive science has created its glorious pedestal…*".[5] It was Antonio Favaro who erected the scholarly monument on which Galileo studies have stood for more than a century. It was called

di Pisa, prevedevano tra gli scopi del nuovo ente, sia la raccolta di pubblicazioni antiche e moderne su Galileo e la sua scuola, sia la promozione e il coordinamento degli studi di storia della scienza relativi all'illustre scienziato e al movimento scientifico che a lui si riconnette. Tali proposte, accolte poi nello Statuto, furono convertite in legge e pubblicate con l'atto di fondazione nella Gazzetta Ufficiale del 22 Ottobre 1941". Cf. the "Storia della Domus dalla fondazione", retrievable at: http://www.domusgalilaeana.it/index.php?id=12

4 Retrievable at: http://www.domusgalilaeana.it/index.php?id=666

5 "Oggi che la scienza positiva ha creato il suo piedistallo glorioso".

Le opere di Galileo Galilei, Edizione Nazionale, Sotto Gli Auspici di Sua Maestà il Re d'Italia.[6]

The *National Edition*

Favaro's *National Edition* of Galileo's works has been the foundation of Galileo studies ever since its completion in 1909. Yet it was a monument erected on the pedestal of positivism—a representation of science that was alien to seventeenth century life-worlds, and which obscured the significance of the fragments that Favaro had collated towards the end of his philological enterprise.[7] The indirect testimonies pointed to the unnerving fact that Galileo had held opinions that he was unwilling or unable to commit to writing. This situation of uncertainty cast doubt on the significance of the books that Galileo did publish. What was their relation to his broader views? Did a loose bundle of doctrines exist that Galileo desired to conceal from public scrutiny? Could such doctrines have something to do with esoteric practices that to Favaro would have looked like pseudo-scientific, if not occult? In this book, I explore Galileo's life-world in search of the hermeticism behind the scenes of his natural philosophical labors. This Galilean hermeticism cannot be equated with the set of doctrines codified in the *Corpus Hermeticum*. Rather it is a form of being-in-the-world that dissolves cultural boundaries, and which tends to manifests itself in the margins of the scientific culture that Galileo inhabited. I will look for his practices of care of the self, his search for the divine, a spiritual goal that is suggested by the symbolism of his late poetry and by his theory of light, on which his pupils and friends collected some of the testimonials that puzzled Favaro.

The Enigmatic Antonio Favaro

Whoever has delved into the ponderous *National Edition* for some time must have remained captive under its spell, feeling the immense spiritual force that brought it to completion. Antonio Favaro devoted half a century to Galileo studies, a good portion of which he published in sundry journals and venues.

---

6 It was promoted by the King of Italy, Umberto I, in 1887, and was to be published at the expense of the Italian state. See Favaro 1888.

7 See Pietro Redondi's fascinating deconstruction of the positivistic images of Galileo in nineteenth-century French and Italian culture (Redondi 1994).

But the *National Edition* was different. In 1907 he published a catalogue of his scholarly works related to it (some 160 publications in total), to which he prefaced an autobiographical "Avvertimento" (*Notice*). It was full of barely disguised pride and almost menacing in its hubris. He wrote: "I do not know when fate will assign to somebody else the task of editing a new edition of Galileo's works, and which criteria will be adopted to produce it. Yet, without immodesty and arrogance, I can state, on behalf of my collaborators and myself, that our successors will have to acknowledge that we proceeded not only with the necessary diligence, but also with such a clear consciousness of our mandate (*alta coscienza del nostro mandato*) that our edition might well be equaled but hardly surpassed".[8]

One is struck by the religious fervor of the phrase "alta coscienza del nostro mandato". There was in that phrase the sense of destiny, of a vocation revealed instantly to him, a religious conversion to which Favaro had been summoned. The master who had worked the miracle of that transformation in the young scholar was Giacomo Zanella, an Italian intellectual and poet whom Favaro addresses as "unequaled master " (*maestro incomparabile*).[9] He tells us that Zanella had shown him a writing by Galileo some fifty years before, but that the thought was far from them on that fatal occasion that Galileo might become the devouring passion of his life. I do not know which work by Galileo Favaro was shown. Perhaps it was *Sidereus nuncius*. Zanella published a book of poems, the first of which was entitled "Milton and Galileo".[10] He treated scientific subjects in his poems not because, he says, the objects of science can be turned into poetry but because scientific discoveries inspire human sentiments. Human beings and their passions, says Zanella, must take center stage, without which poetry is empty.[11] Galileo's sensuous passions beguiled the young Favaro's unconscious, and they became immortalized in the editorial monument—the glorification of the history of science as itself a positive science up against Zanella's poetic evocations of the righteousness of faith, of the *arcanum tremendum*, residing over and above the mundane feats of humans.

---

8 Favaro 1907, p. 4. "Non so quando toccherà ad altri di curare una nuova edizione delle Opere di Galileo, e con quali criteri essa verrà condotta; ma senza immodestia o iattanza parmi poter dire a nome dei miei collaboratori e mio, che i nostri successori dovranno riconoscere, aver noi proceduto, non solo con la necessaria preparazione, ma altresì con così alta coscienza del nostro mandato da poter essere uguagliata, ma difficilmente superata".

9 Favaro 1907, p. 3.

10 Zanella 1868, pp. 1–38.

11 Zanella 1868, p. ix.

That is why, I think, we have fallen under the spell of Favaro's monumental history, pregnant with passion for the human and zeal for philological accuracy and restoration, ourselves seeking deliverance from the sufferings of an age that has seen the most spectacular triumphs of science and technology but also the scientists themselves turning into torturers and villains serving totalitarian regimes. Monumentality bespeaks human passions frozen in space-time and hence become the object of antiquarian contemplation. Favaro's enterprise was the life-long joust of a knight of reason taking up the challenge of the unfinished Italian *Risorgimento*, the separation of Church and State, of science and religion—a malady from which the Italian intellectuals have since never been cured. For, Zanella wondrously portrayed Milton as a *Risorgimento* patriot who had traveled backwards in time to meet—lo and behold—not the puppet of the celebrated fake "*....eppur si muove*" (...and yet it moves) but the antihero who had willingly bowed to the Roman Inquisition. Zanella has Galileo rebut the Protestant Milton by singing the praises of Papal Rome, of the spiritual over the temporal, the triumphant church of the Catholic Reformation resurging after the vain tumults of Luther and his cohorts, and yet not smearing her pristine virtues for meddling with the human all too human. Thus Galileo responds to Milton in Zanella's poem (I offer here a free translation in prose).[12]

"Che fu garzon, che fu? Di tanti moti Qual fin si vide? Dal profondo emerse Roma immortale, che il discisso velo Ricompone longanime, e la preda Con lenta pugna al predator ritoglie. De' Pontefici il fasto, o figlio, assali, E l'immago di Dio scerner ricusi Nel coronato Aronne. Il guardo hai breve, Se dall' ombra scevrar non sai la luce E come il vulgo del parer ti pasci. Visibil sir di non visibil regno, Di Dio la possa e d'uom le colpe ei veste;	What happened, boy? What was the purpose of the tumults? From the depths resurfaced immortal Rome, which restores the torn veil peacefully, and recaptures the prey from the predator following a long battle. You, my son, attack the pomp of the Popes and refuses to see God in the crowned Aaron. You are shortsighted if you cannot discern light from shadow, and you content yourself with popular opinion. Visible lord of an invisible kingdom, He dons the sins of mankind and the power of God.

12 Zanella 1868, p. 15.

Tu nell' uman t'affisi. Ostro e corona Venner co' tempi e dileguar potranno Anco co' tempi: per cangiar di foglie Virtù la trionfale arbor non perde, Perché profonde ha le radici in Dio".	Yet you are stuck with the human. Purple and crown came in time and in time they will go. The triumphant tree does not shed its virtue when shedding the leaves since it is deeply rooted in God.

The Counter-Reformation glories framing the encounter between Galileo's and Milton's souls, so sumptuously invoked by Zanella, their infinite reverberations in the collective consciousness of the Italians, must have captured the imagination of the enigmatic Antonio Favaro.

Zanella's Naturalistic Humanism

Zanella was a master of the naturalistic humanism that had permeated the Italian Renaissance until Galileo and still resonated in Giambattista Vico. Zanella wove together the threads of natural science and the divine project of being. The intriguing poem *Sopra una conchiglia fossile* (On a fossil shell) reveals his cosmic vision of humanity's participation in God's project of universal being. The fossil shell saw the youth of the cosmos while lying on the bottom of a primordial ocean, at a time when mankind had not yet been born (*E l'uomo non era*).[13] The shell witnessed the cosmos ageing over millennia, its vicissitudes slowly changing as the earth underwent its geological transformations. Human beings are young, created a second ago, yet think of themselves as very old. But humans chase the revolutions of time, treading the ashes of centuries while raising their eyes towards the heavens. The mystery of existence is hidden in the darkness of time, past, present, and future, where God has placed the final achievement of mankind. When fate is accomplished, then earth and human will turn into heaven looking forward to God's words indicating the pathway ahead of all times.

Humanity cannot exist outside the natural flow of the divine project of being. Its course is destined to mutate its own substance, from the magmatic mass of early volcanoes, through the forms of life that survived the deluge, eventually metamorphosing into the form of human. But humans desperately pursue nature, emphasizes the humanist Zanella. The poem *Natura e scienza* (Nature and science) speaks of a humanity violating secretive nature, armed

13 Zanella 1868, p. 102.

with instruments and lenses, spying on her back.[14] The reference is to Galileo's telescope. However, spying nature does not reveal to humans the course of the universe, its final destiny, its place in the economy of God's providential plan, and hence nature remains mute. Darkness is not dissipated despite all human endeavor. The substance of humans does not yield to the indiscrete telescope. Their essence lies elsewhere, not definable by naturalistic investigation, but graspable only through the humanistic suspension of the dualism of word and thing. The legacy of Italian humanism is still alive and well in Zanella. Words are not secondary reality, as opposed to *real* things, they evoke the primordial unity of being. The separation of word and thing was effected by the blind rationalism of the Enlightenment, which Giambattista Vico strenuously opposed. Poetry became deviant, imprisoned by the totalitarianism of a scientific reason turned obsessive in pursuit of a silent nature. Life was thus degraded, materialism the new religion, but the humanistic word, perduring in the unprepossessing poetry of Zanella, resisted the violence.

Zanella celebrates the alchemical mystery of life in a poem to the memory of one of his school teachers.[15] Not a vulgar chance game, but pure metal that is rendered liquid by the fire of fate, life could, thus tempered, turn into the gold of God.[16] Life is not mechanism, random collision of a swarm of atoms. It is the transmutation of *prima materia* (prime matter) into the living gold of the hermetic alchemists. It is the hierogamy of suffering and destine, and humanistic poetry is its hierogram. Zanella's poem *Il sonno* (Sleep) sounds the depths of the human unconscious half a century before Freud and Jung.[17] Its beginning is a phenomenology of the transition from wake to sleep, capturing the dazzling passage through the hypnagogic stage. Then images blur into one another, humans swimming together in a river whose waters are not transparent to sunlight. The arrogance of reason fades, and yet tranquility prevails in a world reverberating only with the drama of silence. Auroral light diffuses throughout a scenery where humans do not suffer the boredom of the passing of time. Shadowy dream makes humans equal, the wakefulness of reason defile them into enemies.

14 Zanella 1868, pp. 169–170.

15 Zanella 1868, pp. 194 ff.

16 "A noi la vita non sembrava un gioco/ Vano, volgar; ma di non fral natura/ Nobil metallo che scaldato al foco/ Della sventura,/ Del dolor sotto i colpi e nel lavacro/ De' nostri pianti ritemprato assume/ Gloriose sembianze e simulacro/ Fassi di nume" (Zanella 1868, p. 195).

17 Zanella 1868, pp. 147 ff.

In the inaugural speech of his Italian literature course at the University of Padua, given in 1867, Zanella made a ruthless analysis of the existential dilemma of the natural sciences, on the one hand praising their achievements, but on the other cautioning that positivism erroneously sought to invalidate the spiritual essence of life. From his point of view, physics and astronomy, chemistry and physiology, purged of hypotheses and confused facts, which long hindered their progress, strive to conform themselves to a few simple formulae, so that the laws that govern the universe will become as clear and as rigorous as mathematical axioms.[18] However, the natural sciences, which reduce reality to matter, have blinded humans, obscuring the originary spirit that embraces the realm of the intellect, impervious to the telescope. Positivism defines science as search for the facts of experience. It denies the reality of ideas contenting itself with appearances. Yet humans perceive their being surrounded by an ineffable realm of being, says Zanella, arcane and inaccessible to the senses, but which is nonetheless as certain and as touchable as matter. It is the world of ideas in which spirit contemplates the archetypes of truth, justice, beauty, and order.[19]

Most intriguing was Zanella's defense of philology as a humanistic endeavor, in the footsteps of Leibnitz and Vico. Like a geologist studying volcanoes and fossils the philologist must pursue the history of words scouring the deeper strata of ancient languages for the customs, institutions, ideas and laws of humanity.[20] This is the type of scientific philology, the queen of sciences

18 "La fisica e l'astronomia, la chimica e la fisiologia sgombre dall' ipotesi e dai fatti confusi, che finora le ritardavano nel loro cammino, tentano ridursi ad alcune semplicissime formole: per cui le leggi, che governano l'universo ci saranno messe innanzi coll'evidenza e col rigore di un assioma matematico" (Zanella 1990, I, p. 77).

19 "Peccato che al fiorire di uno studio non sempre rispondano con pari successo gli altri studi! Le scienze naturali hanno talmente invaghito l'uomo dello studio dei corpi, da non lasciargli vedere nell' universo altre forze, che le inerenti all materia: il pensiero, il volere, anzi il principio d'ambedue che è lo spirito, parvero effetti della stessa causa, da cui si formano i corpi. [...] Per quanto una filosofia, detta positiva, siasi studiata di limitare le nostre ricerche ai soli fatti sensibili; per quanto siasi provata di staccare l'attenzione dell' uomo dall' idea per ridurlo a contentarsi dell' apparenza, sente l'uomo intorno a sè qualche cosa di arcano e d'indefinibile, della cui realtà non è meno certo, che della presenza degli oggetti, i quali tocca con mano. Oltre il mondo che noi abitiamo; oltre i mondi che ci sono rivelati dal telescopio, esistono mondi inaccessibili al senso, ma non meno noti allo spirito, che vagheggia in essi l'idee della verità, della giustizia, della bellezza e dell' ordine..." (Zanella 1990, I, pp. 77–78).

20 "Confesso che la filologia in mano ai pedanti è noiosissimo e sterilissimo studio più fatto a spegnere che a vivificare l'intelletto de' giovani; ma se la filologia pigliasse di quell'

(*la filologia riuscirebbe bellissima delle scienze…*), that must have appealed to the young Favaro—an enterprise that was scientific, as the inspiring geological metaphor indicates, but which was nevertheless a humanistic discipline, magnificent and ambitious, whose task it was to excavate the originary languages in search for the words that are not conventional but immediately express the native sense that imitates the object signified, and which attest to the mental powers of the people who first used them.[21] Those who have delved into the painstakingly researched introductions to the volumes of the *National Edition* realize that Favaro's learning, his unbounded ambition to reconstruct Galileo's thoughts by restoring his original words on the basis of original sources, reflected Zanella's humanistic conception of philology.[22] There was in Favaro's soul an unresolved tension between the humanist and the positivist personae.

Even more audaciously Zanella stated that thought and word are not separate, and that thought cannot evolve without the cultivation of the word, and correlatively that word without the substance of thought is empty sonority, brilliance that does not leave durable traces in the history of humanity.[23] There is in his mind an intuition that for all their progress in the investigation of the material universe, the positive science would still owe something much more profound to the care of philologists who perlustrate the places of origination of human history.

A master of naturalistic humanism, Zanella was justly remembered by Favaro. The editor of the *National Edition* was an engineer turned historian of science in the wake of the teachings of his mentor. He came to perceive the

---

ampiezza che il Leibnitz e il Vico le diedero; se […] nell' origine e nella storia de' vocaboli, e quindi nella sentenza che dentro il giro della proposizione rinchiudono, ella si studiasse di scovrire gli antichi istituti, i costumi, le opinioni, i riti, e le leggi del genere umano, la filologia riuscirebbe bellissima delle scienze e feconda di soccorsi mirabili alle altre discipline" (Zanella 1990, I, p. 44).

21 Zanella 1990, I, pp. 44–45.

22 "Alle singole scritture sono premessi avvertimenti d'indole storico-bibliografica, con la indicazione esattissima delle fonti alle quali abbiamo ricorso, dei criteri che abbiamo seguiti nel servircerne e delle convenzioni che si sono stimate opportune per rappresentare, più esattamente e più semplicemente che fosse possibile, il risultato del nostro lavoro" (Favaro 1899, p. 130).

23 "Funesto in ogni tempo alla letteratura italiana è tornato il pregiudicio di credere che pensiero e parola siano cose dverse, le quali separatamente si possano coltivare con gloria. Il pensiero né si svolge né si matura senza l'aiuto della parola, per cui vestito di forme sensibili esce dall' intelletto e passa ad operare sull' animo altrui; e d'altra parte la parola, senza la sostanza del pensiero, che sia come corpo alla veste, è vuota sonorità, brillamento che fugge, né lascia traccia durevole sulla via dell' umano progresso" (Zanella 1990, I, pp. 43).

great paradox of the historical facticity of the natural sciences, their belonging in both nature and soul, which resists integration in the materialistic world-view. And yet as a true apostle he preached the deeds of the founding father of the experimental method. Humans, warned Zanella, cannot be defined by the natural sciences. Thinking escapes the purview of the instrumentalities excogitated by the cunning of reason. Materialism is a preposterous metaphysics, and life a dreamy transmutation into divine spirit.

The *Codice Galileiano*

The *Codice Galileiano* (Galilean Code) was Favaro's uncompromising response to Zanella's plea for tempering history's posture as positive science with the living tradition of Italian humanism. In 1907, the nineteenth volume of the *National Edition* was published. It contained the *Galilean Code*. Favaro proudly announced in the introduction that the volume was more than the accomplishment of a better edition of Galileo's writings. It was a *code*, "a systematic collection or digest of laws; a body of laws so arranged as to avoid inconsistency and overlapping" (*Oxford English Dictionary*). The OED definition captures impeccably the intent that must have guided Favaro in compiling the volume.[24] It was to give a normative collection of documents that would once and for all establish the reputation of Galileo as the hero who founded the religion of the experimental method. For, all the mystery of this cult of the document is really concealed in the meaning of the word 'document', which represented the incarnation of the positive spirit that initiated the history of science as an empirical science in its own right.

Favaro divided the documents in forty-five headings, most of them having further subdivisions. We are not told by Favaro what constitutes a document. But he indicates that some materials in the volume are not to be considered documents, and that at most they count as quasi-documents. These are biographical accounts written by Galileo's contemporaries.[25] The choice of these ambiguous materials was made following loose criteria that tended to select either manuscript materials or writings by ocular witnesses of the events (...*spettatori stessi degli avvenimenti*).[26] Hence we can conclude that ocularity

24 "...fin adesso nulla s'è avuto mai che corrisponda al concetto di questo che possiamo veramente chamare Codice Galileiano" (Galilei 1890–1909, XIX, p. 7).

25 Galilei 1890–1909, p. 9.

26 Galilei 1890–1909, p. 9.

played the fundamental role in steering Favaro towards canonizing the sparse biographical material in regard to the *Galilean Code*. Among other things he excluded a large number of poems by Galileo's contemporaries that he disqualified as eulogistic (*encomiastiche*), adducing the disclaimer that some of them had already been published in previous volumes of the *National Edition*. How does one, then, circumscribe the events, or facts, that constitute a human life? What counts as an authoritative document, or a historical fact? The root of the word *document* is the Latin *docere*, to teach, hence what constitutes a document should not rule out anything that might be instructive. The historiographer worshipping the naked document, whatever that may be, became the sickly sibling of the natural scientist idolatrizing the experimental fact. Yet anything that educates us about the human—indeed the plenum of life itself as embracing the whole of humanity and the cosmos—and that transcends the sparse ontology of facts admitted by positivism should be recovered and restored.

As Favaro knew well, Galileo himself planned on publishing a second edition of *Sidereus nuncius*, this time in Italian, accompanied by poems that he acknowledged had already entered the tradition of his astronomical discoveries.[27] What positivist dogma could not tolerate was the idea, implicit in Galileo's intentionality, that scientific facts are not restricted to empirical procedures for the manipulation of apparatus, their mythical place of origination, or to the supposedly authoritative account given by the first manipulator. Even less acceptable to positivism is the idea that collective consciousness, or the unconscious psyche, are constitutive of the facts—including consciousness's literary modalities, which are perceived as separate from other modalities of thought by positivists of all persuasions. In early 1610, the manifold of formative realities that constituted Galileo as a human being living within a memorial tradition was already radiating from his consciousness.

It was Nunzio Vaccalluzzo that rescued from oblivion the poetic tradition that had formed around Galileo, thus challenging the positivist renegation.[28] Vaccalluzzo called into question Favaro's dogmatic attitude in the compilation of the *Galilean Code*. He claimed that the poems, though artistically modest, did have both historiographic and biographical value and as such must be

27 "Questa seconda volta credo che lo farò in lingua toscana, sì perchè, oltre a i librai, ne son pregato da molti altri, sì ancora perchè credo che le Muse toscane non taceranno in così grande occasione le glorie di questa Ser.[ma] Casa, perchè sin qua sono alcuni che scrivono in questo proposito: et tali componimenti si potranno prefigere all' opera" (Galilei 1890–1909, X, p. 300).

28 Vaccalluzzo 1910.

counted as documents of the scientific movement promoted by Galileo.[29] Disappointingly, Vaccalluzzo, infatuated with the Galileo myth and with nationalistic sentiments, did not really know what to do with the poetic materials rejected by Favaro.[30] Perhaps his sharp intellect was blinded by the literary prejudice, still shared by many, that poetry in seventeenth-century Italy had degenerated into a dead palsy of the soul. He ended up concocting a tabloid story—almost a sociology of science in miniature—that portrays Galileo as a shrewd self-promoter, canvassing support for his telescopic discoveries from encomiastic poets, with an eye to improve his career by seeking employment back to the Court of the Grand Duke of Tuscany.[31] Aesthetic canons stood in the way, as they still do. More than a century later, a valiant historian, John Heilbron, swayed by positivist dogmatism, repeats commonplaces such as the following. "The poets were to serve Galileo as newspapers do modern scientists, advertising and celebrating discoveries before the experts had their say".[32] The rehabilitation of the poetic corpus that Favaro had discarded remains a task for the future.

The Dead Palsy of the Soul

The seventeenth century went down in Italian literary history as a period of poetic degeneration, a disintegration of the Italian collective consciousness, in sharp contrast to the almost unanimous applause to the scientific movement

29 "Vero è che brutta rinomanza ha, in genere, la poesia del seicento e bruttissima, in ispecie, la poesia encomiastica e cortigianesca; ma, presi nel loro insieme e indipendentemente dal loro valore artistico, questi componimenti hanno un valore storico e biografico non disprezzabile, come *documento* di quel secolo e di quel movimento scientifico promosso da Galileo" (Vaccalluzzo 1910, p. xxi).

30 He refers to a letter by Tommaso Campanella to Galileo as being "magnifica d'empito poetico e di nazionale fierezza" (Vaccalluzzo 1910, p. xxv). Vaccaluzzo's work was published under the watching auspices of Favaro, as we read in the *Acknowledgments* (Vaccalluzzo 1910, p. lxxv).

31 Vaccalluzzo 1910, pp. xxvi ff.

32 Cf. Heilbron 2010, pp. 165–166, who has little patience for the 'advertising industry' of the seventeenth century. Judging by the 'verses' commissioned by Oxford University Press to improve sales of Heilbron's book ("A masterpiece…It far surpasses all previous biographies of Galileo. Impeccable scholarship", by Nick Jardine, and "By far the best general reconstruction of Galileo's private and intellectual life available in the English language", by Paolo Galluzzi), change in matters of encomiastic style has not been for the better (see back cover of the book for the quotes).

inaugurated by the glorious Galileo. This negative opinion was shared by the vast majority of literary critics, intellectuals, and writers of the subsequent two centuries. It persisted well into the late twentieth century—still inculcated in the minds of my generation by scores of professors who recycled clichés established by national curricula. Vaccaluzzo, as we have seen, was no exception. But it remains unclear what the motivation was of the vehement reaction that ended up paralyzing intellects of the caliber of Benedetto Croce and Antonio Favaro. It certainly changed over time, at least in terms of the patterns of explanation that were brought to bear on the controversy. The insurgence against the 'emptiness' of seventeenth-century poetry, its being devoid of… something that was rarely specified, appears to have been the most frequently repeated allegation.

Benedetto Croce argued that the literary decadence of the seventeenth century found its theoretical justification in the theory of *ingenium*. It was opposed to the faculty of intellect under the influence of ancient rhetoric, which instead of being based on true reasons could exploit probability, verisimilitude, enthymemes and examples, in place of conclusive syllogisms and induction. The *vaniloquium* (vain speech) of seventeenth-century literary decadence, says Croce, found its own theoeretical justification in this rhetorical theory. Prose and verse did not want to show the true but the curious, the superficially clever, the ingenious.[33] Croce noted that throughout the seventeenth century the poet's imagination was praised for being the organ of a phenomenological suspension of the values of reality that did not pretend to portray truth but only encode perceptions.[34]

It was Giambattista Marino's magnificent *Adone* (first published in 1623, the same year as Galileo's *Il saggiatore*, consisting of twenty *canti* in octaves totaling more verses than Ariosto's *Orlando furioso*) that came to epitomize the decadence. Corrado Corradino called it an *interminabile poema* (interminable poem) that, however, received universal and enthusiastic approval from the contemporaries. Though he concurred with the negative opinion, Corradino acknowledged that Marino's poetic achievement reflected a universal trait of consciousness, and as such would have to be taken into account

---

33 "Il vaniloquio della decadenza letteraria secentistica trovava in questa teoria rettorica la sua giustificazione: quelle prose e quei versi, marineschi e achillineschi, non volevano esibire il vero, ma l' appariscente, l'arguto, il curioso, l' *ingegnoso*" (Croce 1902, p. 193).

34 Croce 1902, p. 198.

as a most genuine manifestation of a historically determined form of life.[35] For Corradino the reason for the abject decadence was a failure of political self-consciousness, namely, the Italian servile resignation to Spanish rule. The enslaved individuals, their liberty negated by the historical drama, sought solace in degenerate poetry. Since liberty could not be enjoyed in freedom of thought one looked for an ersatz satisfaction in dehydrated formalism.[36]

The bill of indictment was summed up by the literary prosecutors with the word *concettismo*. Croce wrote a ground-breaking essay on *concettismo* in which he explained that from Aristotle the seventeenth-century theorists, essentially Aristotelian in outlook, had borrowed a reverence for metaphor as a means of learning through pleasure. Hence their reasoning was impeccable. It was as follows. Let us pile up and complicate metaphors, and in this way we will increase pleasure and learning.[37] Not that there were lacking critics who came to oppose *concettismo*. Matteo Pellegrini, Croce points out, was the most acute of the dissidents. But he could not follow through his insight into the correct theory of literary form (according to Croce's idealist orientation) as the powerful vehicle of thought and not its ornament.[38] However, Croce's insistence on *concettismo*'s desecration of the union of form and matter that should be constitutive of poetic expression—in accord with his idealistic principles — is marred by anachronism. For one thing it does not shed light on why

35 "Vuol dire che questa opera è un riflesso fedele della coscienza universale la quale vi si trova ritratta in tutto o nella sua parte più viva; vuol dire che essa è interprete sicura di quel particolare concetto della vita che hanno gli uomini in un dato periodo del tempo e che varia continuamente secondo le conquiste del pensiero; o nella peggiore delle ipotesi, vuol dire ch'essa corrisponde almeno a quelle generali tendenze le quali in certe epoche, per forza di circostanze diverse, trascinano in tal modo gli ingegni e gli affetti delle generazioni, da confondersi quasi colle manifestazioni più *schiette* della vita propria e vera"(Corradino 1880, p. 2).

36 "Nei primi anni del secolo decimosettimo, quando l' Italia rallegrava i suoi ozi servili colla lettura di versi pregni di mollezze e di allegorie lascive; quando da tutti si cercava di soddisfare il bisogno prepotente di novità col cercarla non nel pensiero, che non aveva libertà, ma nella forma e nella frase..." (Corradino 1880, p. 1).

37 "Così ragionavano i seicentisti, quando ragionavano; e bisogna convenire che la logica era dalla loro parte" (Croce 1899, p. 2).

38 "In queste poche righe è accennata la teoria moderna della forma letteraria, considerat non più come *ornamento*, ma come il *veicolo* del pensiero. La forma che non si fa notar come cosa indipendente, e pure serve di mezzo a comunicare possentemente il pensiero, è la vera forma estetica perfettamente trasparente. Ma un pensatore di quel tempo non era preparato ad accogliere e fecondare questo prezioso seme, nel quale s'imbatteva come per caso" (Croce 1899, pp. 16–17).

seventeenth-century poetry, for all its faults, could look deep into the abyss of emptiness. The daring gaze neutralized the vertigo of the infinite, which stirred up both art and science, and thus elevated the gyrations of *concettismo* to the rank of supreme harmony of emptiness. The universe had been voided. Giordano Bruno and Galileo had accomplished the voiding of the millennial spheres. The telescope interposed multiple glasses between the gaze and the object of vision, like the metaphors of poets, estranging the object become imaginary from its material substratum.

What Croce failed to see was that an optical tube could effect the endless multiplication of images, progressively separating the material substratum of the sensory object from the image formed by the organ of sense. The desecration of the union of form and matter, of consciousness and sensory object, which disrupted the assimilation of the visual species into mind, according to then current theory of vision, was already implicit in Galileo's ingenious adaptation of the play of lenses to astronomical observation. It was a violent rupture of planes that ushered in the age of astronomical *concettismo*. The language of astronomy became involved in optical metaphors that sublimated the material substratum of sensory objects into fantastic images. *Concettismo* became the intellectual scaffolding of vision within Galileo's life-world.

What else if not *ingenium* could have brought the *perspicillum* (Galileo's early word for the yet not invented *telescope*) into existence? Certainly not the strange rumors of optical instruments, reaching Padua from the Netherlands, honestly mentioned by Galileo. *Concettismo* underpins the rhetorical strategy of his announcement of the magical optical tube in *Sidereus nuncius*. The convolution of the wording of the book's frontispiece was a device of *concettismo* masterfully calculated to leave the reader in a state of suspended animation. No *secentista* could have done better. The spectacles contemplated by Galileo with the help of a *perspicillum* just discovered by him,...*PERSPICILLI nuper a se reperti beneficio*, as we read in the carefully typeset frontispiece, were spectacles of the poetic imagination unleashed by the dissociative power of the instrument. The optical tube worked like a hallucinogen in the early seventeenth century. Archetypal realities emerged from the collective unconscious that had been excited by the creative dissociations of consciousness realizable through the visions of *concettismo*.

Then Galileo explained the multiplication of visual objects as they appear through the *perspicillum*. He had excogitated a practical procedure to estimate the apparent equality of two plane figures, paper squares or circles, as seen at the same time with one naked eye and the other close to the eyepiece of the instrument.[39] The optically multiplied object became a metaphor of the sensory object, visible or invisible with the naked eye, as the case might be. Multiplication was potentially infinite since it could transform the invisible into the visible—an optical model of *creatio ex nihilo*. This was the case of the moons of Jupiter, invisible with the naked eye, named by Galileo the 'Medicean stars'. The abyss of the universe could from now onwards be scrutinized inspecting the poetic detritus of the disintegration of consciousness triggered by the *perspicillum*.

Copernicanism and the Imagination

It was Lodovico Antonio Muratori, at the end of the seventeenth century, who came closest to confronting the root causes of the distress brought about by *concettismo*. He severely censored the palsy of the soul that, in his view, had strained the nerve of Italian poetry. For Muratori the paralysis was due to abuse of figurative language. What, according to him, *concettismo* was guilty of consisted in this. Once a metaphor has been formed correctly on a ground object, the poet cannot automatically attribute the qualities and operations of the ground to the translated object, for the translation of the name from one object to another does not authorize the translation of all the effects, properties, and natural qualities of the ground object. For this would imply that the translated object would exceed its own being, which belongs wholly in the realm of the imagination, and that it would acquire an ambivalent existence, a novel being in competition with the real object, whose name has been transported to it.[40] Muratori probed deep into the nature of *concettismo*. The play of translations must terminate before a transmutation of beings occurs. He offers the following example.[41] There is no doubt that there is a similarity between fire and love. Thus it is legitimate to form the metaphor 'love is a fire'. But it

39 Galilei 1610, pp. 6 verso–7 recto.

40 "La terza osservazione, che è forse la più necessaria, si è: Che quando la fantasia averà formata qualche immagine, o traslazione con giusto fondamento, non potrà poscia il poeta, se non scioccamente, attribuire all' oggetto metaforico, o traslato, le altre operazioni, e qualità dell' oggetto proprio, quasi che per essersi trasferito il nome di una cosa ad un'altra, fosse lecito anche il trasferire ogni suo effetto, ogni sua proprietà, e qualità naturale; o come se la traslazione più non fusse immagine della fantasia, ma l'oggetto vero, di cui s'è trasferito il vocabolo. E questo è l'errore, over per l'ordinario cadono gli amadori de' concetti falsi, per isperanza di cagionar meraviglia in chi legge" (Muratori 1730, p. 289).

41 Muratori 1730, p. 289.

would be nonsensical to translate all the properties and qualities of real fire to the imaginary fire of love. It would be silly to say that love can dry, burn, and be extinguished with water, all of which naturally occurs with real fire. Muratori castigated Marino for forming an exorbitant metaphor in order to eulogize a Flemish painter (Peter Paul Rubens) who had vividly represented the fall of Phaeton. Marino's metaphor says that if the painter had given life to fire, as he had brilliantly done to the young boy (Phaeton), then it would have burned not only the canvass but the whole world.[42] But Muratori insisted that although it is legitimate to praise an excellent painter by saying that he gives life to objects on canvass, yet one cannot attribute to imaginary life all the effects and operations of real life.[43] Imaginary fire does not burn. But what was the discomfort all about?

To my mind Muratori missed one element in the workings of imagination that had changed over the course of the seventeenth century. The geocentric cosmos of antiquity had been dismantled. The mighty sun took center stage. Thus astronomy changed the imagination. The motionless earth at the center of the cosmos had been removed. It became groundless, floating in empty space. The imagination in turn became groundless, free-floating in consciousness, or rather hovering over the borderline (if one exists) between the conscious and the unconscious. Reality lost its earthly ground. Marino's imagination soars into unheard of universes where painted suns do acquire real life. The form of life we experience in this universe is only one of multiple experiences open to the Copernican imagination. *Concettismo* was not a palsy of the soul but glossolalia of the novel imagination, a voice speaking in the wilderness, reporting about the transmutation of beings in the abyss of consciousness.

Glossolalia

The iridescences manifesting themselves through the *perspicillum* transformed Galileo's language. He responded to the cosmic visions in a preaching modality, as though he had received the supernatural gift of glossolalia. He admitted this much explicitly invoking divine grace—*divina prius illuminante gratia*.[44]

---

42 "Che se, come al Garzon la vita avessi/ Dato alla fiamma: ancora di nuovo avrebbe/ Non che le tele, incenerito il Mondo" (Quoted in Muratori 1730, p. 290).

43 Muratori 1730, p. 290.

44 "All these things were observed and discovered a few days ago with the help of the *perspicillum* devised by me thanks to the preceding and illuminating grace of God (Quae

He used religious language uttering words like an oracle—"we were first impelled…to admonish" (*ad monitos faciendos…nos apprime impulit*).[45] The language he could now speak for the first time, voluptuously and visionary, was Latin. Until 1610 Galileo wrote little in Latin, though his competence was growing fast. Most commentators (who do not write or speak Latin) repeat the hackney cliché of Galileo's poor Latin.[46] This is puzzling since it is hard to make a judgment about the qualities of a piece of writing in a language that one does no longer speak or write any longer. In the case of *Sidereus nuncius* one has to enter into its labyrinth of linguistic grotesqueries, learning how to feel the ecstatic voice, the spattering mouth speaking in a tongue occasionally resounding with biblical echoes. Clifford Truesdell remarked that "[t]he workshop of the scholar in the history of science is the periods in which his authors lived. He should know those periods' ways of life and belief and education, both the common and the eccentric; their political histories; their variety in aspects; their social and economic structures, their architectures, literatures and arts. He should feel at home in the houses of those times, sit easily in their chairs, both figurative and wooden, and discern what was then mostly admired or rejected in painting and sculpture and decoration. He should have read not only the books that carried the intellectual products of his period but also those that were then the fare of young minds as they were taught, such books having been commonly of an earlier time. The student who does not command, as a minimum, the main episodes of Holy Scripture, classic mythology, and the corpus of Golden Latin is glaucomatose in the modes of thought of Western men educated before 1900".[47] This is the manifesto that I ideally subscribe to in order to prevent philosophical glaucoma, and thus restore my sense of at-homeness within Galileo's life-world.

Sidereus nuncius begins by simply intimating that great things will be spoken of, that it is great to add many more stars to those long known, and that it is most beautiful and most funny to see the body of the moon

---

omnia ope perspicilli a me excogitati divina prius illuminante gratia, paucis abhinc diebus reperta, atque observata fuerunt)" (Galilei 1610, p. 5 verso).

45 Galilei 1610, p. 5 verso.

46 Cf. Francesca Romana Berno's important essay on Galileo's Latin (Berno 2008), which furnishes a balanced assessment of his mastering its rhetorical resources in a variety of different contexts.

47 Truesdell 1976, pp. 631–632. Truesdell has portrayed at-homeness beautifully. The scholar menaced by professional disciplinary boundaries is never at home. He is under house arrest.

close-up.[48] It was quite something, says Galileo, to demolish a bunch of controversies about the Milky Way, a piece of cake, showing its essence to the senses and to the intellect, and most funny of all was to plainly demonstrate the substance of the nebulae to be rather different than hitherto believed. After all this entertainment, tumbling down from heaven like a torrent of laughter at human gullibility, comes the prophetic moment. Now Galileo preaches to astronomers and philosophers that four stars never seen before orbit one of the known planets. Passing from laughter to prophecy takes no time in the ecstatic state of mind; in early modern Latin, just a *Verum*, with which Galileo's tone shifts to the modality of prophetic admonition, before uttering the punch-line evoking the illuminating grace of God.[49]

The preternatural gift of glossolalia finds its natural means of expression in the style of purest *concettismo*. The incantations of the moon performed by Galileo under the influence of the *perspicillum* are the hyperbolic masterpiece of *concettismo* in the whole *Sidereus nuncius* (together perhaps with the calculation of the height of lunar mountains). They are reported in the book in the vicious form of the extended metaphor that was decried by Muratori against the exorbitances of Marino. Aided by the magical *perspicillum* Galileo has divined countless maculae on the surface of the moon, which are not visible with the naked eye, and which he calls appearances (*apparentiae*).[50] These divine appearances are shifting patterns of light and darkness whose rhythm of variation harmonizes with the changing position of the sun. This rhythm, Galileo recounts, is analogous to the mundane rhythm of illumination that we observe in our valleys here on earth when the sun rises. Now, let the bold analogy be extended secundum artem. Then it will become wholly

48 For contextual information about *Sidereus nuncius* (such as the meaning of the title, optical theory and lenses, Galileo's early career, and the events closely connected with the publication of the book), cf. Isabelle Pantin's edition (Galilei 1992), with extensive introduction, notes, and bibliography. Her commentary has remained the most informative to date.

49 Galilei 1610, p. 5 verso. The *Verum* starts a new paragraph in the book but in general paragraphs were not marked by beginnings on new lines in the original manuscript, for the latter was like a flow of lava with no demarcations of passages. Cf. the Ms. Gal 48, preserved in National Library at Florence, or the facsimile in Galilei 1890–1909, III.1, pp. 17 ff. I am puzzled by the fact that Favaro did not content himself with collating the text of *Sidereus nuncius* based on Ms. Gal. 48, an editorial practice that he would follow for almost all texts throughout the *National Edition* (Galilei 1890–1909), but decided to publish the facsimile of the manuscript. Perhaps he realized that the modalities of thought reflected in the flow of the manuscript could not be congealed into a definitive printed text.

50 Galilei 1610, p. 7 verso.

understandable, claims Galileo, that the surface of the moon is not polished, even, and perfectly spherical, as philosophers have thought, but that on the contrary, the moon is uneven, jugged, full of cavities and excrescences, exactly like the surface of the earth, which is rigged with high peaks and deep valleys here and there.[51]

The religious and magical motives associated with the *perspicillum*, traces of the hermeticism I am looking for, may not be immediately evident to the casual reader of modern translations who has lost the heritage of classical culture, and above all does not read Latin fluently enough to have a feeling for Galileo's prose and diagnose the linguistic oddities of the divinely inspired writer. Galileo's glossolalia speaks the language of the Vulgate but also exploits the rhetorical resources of *concettismo*, a veritable *coincidentia oppositorum*, reporting imaginations of the heavens that transcend common parlance. It was a cognitive resource fully operative in his life-world that we must still learn how to appreciate.

The objective spirit of institutions designs the style of inhabitance of the individual within a culture, the possibilities of her or his individuation within the matrix of a historically given tradition. The phenomenon of *concettismo* disinhibits an array of modalities of speech in the multilingual context of the Italian Renaissance. The pursuit of extended metaphors encouraged by its style opens up a window for the imagination to explore new vistas, revealing universes that can be imagined with the help of the senses in concert with magical instrumentalities. So the individual immersed in a life-world where *concettismo* is an institution, indeed being at one with it, does not feel the constraints of reality that became the norm of perception in the age of science. Being-in-a-life-world does not follow rules of existence established by scientific method. It is a being-for-ife in its panoply of manifestations. Reality is a construct of being-in-a-life-world, and depending on the institutions forming the life-world multiple realities are open to the imagination to be asserted into existence. This dissociation of reality, or multiplication, is not accessible from the perspective of positivist historiographies. These are bound to mimic the methodic rules of the positive sciences, materialism being their tacit postulate, and hence to be subservient to an impoverished metaphysics.

The institution of *concettismo* animating Galileo's life-world suggests that the positivist interpretation of his so-called telescopic discoveries needs to be reconstructed. The facts laid bare by the discoveries made with the *perspicillum* cannot be sharply differentiated from the poetic madness that Galileo

51 Galilei 1610, pp. 7 verso–8 recto.

himself cultivated as a form of life. A continuum of human experience underlies the sparse ontology of empirical facts that positivism is at pains to isolate on the surface of history.

The Attunement of Grace

Galileo's evocation of the prior operation of the illuminating grace of God in the work of cognition—...*divina prius illuminante gratia*—has never been taken into account in the literature (as far as I have been able to ascertain). The prevailing materialist prejudice in contemporary academic philosophies has prevented scholars from responding to the claim that Galileo makes in this crucial passage of *Sidereus nuncius*. To place ourselves in a better position to answer this claim, we need to dissolve the boundaries that separate ours from Galileo's life-world. These boundaries mostly concern the loose bundle of beliefs that orientate modern conceptions of humans as sharply distinct from the non-human other, be that the divine, or the transcendental, or the realm of non-human animals.

The frontier that needs to be renegotiated is the meaning of the definition of humans as *rational* animals that differentiates the history and philosophy of science from other academic disciplines. It establishes a sharp differentiation between the human and the non-human by demarcating a region of reason from a territory of non-reason without asserting anything more specific about the character of this territory. In the wake of modern science, this definition has acquired the tacit implication that scientific reasoning, or rationality, whatever that may be, sets humans apart from other forms of being. This setting apart of us as rational, as opposed to the a-rational other, is deeply ingrained in the life-world which we inhabit. In contrast, within Galileo's life-world there is no sharp demarcation of the faculties that define human rationality. This is not to say that in his time humans, for instance, were not marked off from the beings of the animal and plant kingdoms according to a similar definition emphasizing the rational soul. But the fact is that reason as an attribute of soul was inclusive of, rather than antagonist to, faculties that we would not call rational. Thus the work of illuminating grace was intertwined with the rational faculties of human beings. We may call the state of mind-and-body, or state of being-in-the-world, induced by the prior operation of the illuminating grace of God, an *attunement*. Then, what Galileo is pointing to as a prerequisite of cognition, as the hidden work of revelation that is operative when being-in-the-world is mediated by the *perspicillum*, will appear in the light of this attunement of grace.

Because of the materialist prejudice the work of grace has unconsciously been suppressed by academics who have studied *Sidereus nuncius*. Possibly an object of inquiry for historians of religion, it cannot be the prerogative of historians and philosophers of science. However, since materialism is not a scientific fact but rather a scientifically indemonstrable metaphysics, it remains unclear how to answer the claim made by the evocation of grace in the processes narrated by the sidereal messenger. To open up the question of the attunement of grace in relation to *Sidereus nuncius*—a difficult question that at any rate cannot be settled here—I will cursorily explore the theological thinking regarding illuminating grace in the seventeenth century. I do not intend to survey a field as vast and as complex as the debates about grace in early modern theologies, for that would be like writing a fundamental chapter of the religious, political, and intellectual history of Europe.[52] In fact I will restrict myself to one illustrative work on grace, by Gabriele da San Vincenzo, and simply point to a few relevant motives that I wish to weave into the fabric of *Sidereus nuncius*.[53] To achieve a measure of clarity it is imperative that an effort be made to suspend, or bracket, for the time being, the validity of the materialistic framework that defines human cognition as some sort of mind-body relationship, either in terms of reductionist biochemistry, i.e., thinking equated with neurophysiology, or in terms of consciousness as an object of empirical psychology.

Human beings live in a state of corruption because of original sin. There are epistemic consequences that originate with that transgression. Can human beings attain knowledge of truth in their fallen state? Gabriele da San Vincenzo answers by distinguishing speculative truth considered analytically from speculative truth considered synthetically.[54] Here speculative is contrasted to practical in the sense of knowledge of action to be undertaken. Speculative truth is a truth, or ratio (*proportio*), which our intellect can know through the medium of the senses. It is called natural truth both on part of the object [*ex parte obiecti*] and on that of the medium (*ex parte medii*). There are natural

52 See Feldhay 1995, who reconstructs these debates in relation to Galileo's encounter with the Roman Catholic Church, which led to his prosecution and infamous recantation in 1633.

53 Cf. Brocchieri 2012, pp. 175 ff., on Gabriele da San Vincenzo's career.

54 "…verum speculativum divisive sumptum" and "…verum speculativum collective sumptum", in Gabriele da San Vincenzo 1658, p 36, and 41, respectively.

truths, however, that are not accessible to knowledge, so they are not such *ex parte medii*, since they cannot be mediated by the senses. These unattainable natural truths are, for example, the occult thoughts of the soul (*occultas cordium cogitationes*), or future contingents, or the number of grains of sand in a sea. For them there is no natural middle term that would afford a connection between a subject and a predicate. The natural sciences, Gabriele da San Vincenzo points out, are all good, and they can be put to good use.[55] Now, since it is clear that the sciences teach most evident truths, the question is whether the speculative intellect of humans in their fallen state can still attain the truths of the sciences with the help of grace. Gabriele da San Vincenzo's opinion is that humans in the fallen state can attain speculative truths *analytically* but only with the general help of grace.[56]

But what about the totality of speculative truths, are they attainable *synthetically*? In this regard Gabriele da San Vincenzo states that the consensus among theologians seems to be that, even in their fallen state, human beings have preserved a physical faculty (*potentia physica*), a power commensurate, proportional, to its proper intended object, by which they can attain knowledge of all speculative truths synthetically.[57] But for humans to attain perfect knowledge of a science such as logic or physics or metaphysics, the help of a very different type of grace is required.[58] This is the illuminating grace [*gratia illustrationis, et illuminationis*], which operates in tandem with an affective grace [*gratia piae affectionis*] that will strengthen the will and fortify the intellect in order to persevere and thus penetrate deeper into the speculative truths.[59]

The question can be raised, then, about what type of work is done by the divine assistance of grace that becomes operative when being-in-the-world, in the thick of the process of inquiry about natural truths, is mediated by the magical *perspicillum*. In the light of the attunement of illuminating grace—recall Galileo's explicit evocation in *Sidereus nuncius*—sensory observation through the magnifying glasses of the *perspicillum* takes the form of a disillusionment of the intellect. The speculative intellect, unaided by the instrumentalities

55 Gabriele da San Vincenzo 1658, p 36.

56 Gabriele da San Vincenzo 1658, p 37.

57 Gabriele da San Vincenzo 1658, p 41.

58 Gabriele da San Vincenzo 1658, p 42.

59 "Nostra sententia sequenti explicatur conclusione. Ad hoc ut intellectus speculativas omnes percipiat veritates collective, ultra gratiam externae protectionis etiam gratiam requiritur internae illustrationis ex parte intellectus, et ex parte voluntatis requiritur gratia piae affectionis" (Gabriele da San Vincenzo 1658, p 44).

of the senses, is obfuscated by a curtain of ignorance. Lifting the curtain is the work done by the divine assistance of grace that becomes operative when being-in-the-world finds itself in the thick of a process of inquiry about natural truths with a *perspicillum*. This lifting of the curtain is not automatically done by the magical instrument. Because you are in a state of corruption you see nothing with it unless illuminating grace intervenes by lifting the veil. In Galileo's life-world the instrumentalities of the senses do not function on the same principles as modern scientific apparatus. The use of the word *telescope* in the scholarly literature, though the Italian *telescopio* was legitimated very early on, has been so misleading as to render formative realities, objectively present in Galileo's life-world, virtually unintelligible.

The sidereal messenger announces great and most admirable spectacles that become visible once the veil obfuscating the intellect has been lifted. To reinforce the disillusioning work of grace, being-in-the-world, while in the thick of the process of inquiry about the totality of natural truths, finds an instrument that entices the species of objects to enter the sensory organ of vision. The universe above the sphere of fire, towards whose objects the *perspicillum* can be trained, enfolds a totality of natural truths that can be unfolded by the human intellect illuminated by divine grace. The sidereal messenger states that it is quite possible that in the future more valid things will be uncovered by others, or by the writer himself, with the *perspicillum*, or with a similar device.[60] It remains an open possibility for fallen humans, aided by novel instrumentalities of the senses, to encounter along their pathways the gratuitous work of grace illuminating their processes of inquiry about natural truths.

But Galileo also stresses that it was thanks to his immense labors that he finally stumbled upon the most exquisite *perspicillum*, the one through which the gates of heaven would finally be flung open. The Herculean labors that perfect the optical instrument have conjured illuminating grace to lift the veil of the intellect. Hence he wishes to remember not only its construction but the event that occasioned its discovery.[61] In order to commemorate the labors, he memorializes the historic events by telling a myth of origins. Small wonder that he planned on incorporating in the second edition of *Sidereus nuncius* the poems that the *secentisti* wrote to immortalize the hero. Yet the intervention

---

60 "Alia forte paestantiora, vel a me, vel ab aliis indies adinvenientur consimilis organi beneficio…" (Galilei 1610, p. 5 verso).

61 "…cuius formam, et apparatum, necnon illius excogitandi occasionem prius breviter commemorabo" (Galilei 1610, p. 5 recto).

of illuminating grace appears to have been solicited by an essentially magical procedure. The polishing of the lenses, the grinding of raw material, the nocturnal and fiery images of the flamboyant glass-blowers in Venice, and the shaping of the cylindrical tube, all of which constitute human labor in as many strange facets, become at the hands of Galileo, the Herculean laborer, a magical procedure. It was the operation of attracting virtues from the heavens recommended by the Florentine magician Marsilio Ficino. Galileo the magician of glasswork hoped to receive the attunement of illuminating grace. The totality of truths, which our intellect can know through the medium of the senses, when prior illuminating grace intervenes, becomes accessible through the labors that perfect our instrumentalities of the senses. We become the living totality of truths when we receive the gift of illuminating grace. But the labors that perfect our instrumentalities of the senses entice divine grace. This is a daring possibility hinted at by the sidereal messenger. We are not fallen beings in the sense that our being-in-the-world is irremediably flawed—this would be the full-blown protestant heresy. We participate in the work of salvation that divine grace explicates through our labors. We are partners in the project of being that God launched with the labors of the six days of creation. This motive of partnership with the divine, the completion of the divine through the human, is a Hermetic theme that finds an echo in the account of the starry messenger. It does challenge Catholic orthodoxy as defined by the Council of Trent.

Sacramental Poetry

The reaction against the Italian *seicento* and *concettismo* that had gathered momentum throughout the eighteenth and nineteenth centuries was the compensatory backfiring of the European collective unconscious, which became constellated by the enforcement of the mechanistic and materialistic dogmas of modern natural science. The latter had excluded the soul from the domain of scientific inquiry—illegitimatizing *res cogitans* as an object of rational pursuit while sanctioning the empirically accessible laws of *res extensa*, its soulless bodies and the physiological correlates of perception. Mechanism displaced spirit, and the mythology of scientific fact displaced poetic myth. The void was not in the convolutions of seventeenth-century poetry, or in the vortexes which replenished the universes of the poetic mind. It was modern consciousness that projected its loss into the plenum of the imagination. The poetic mind of the seventeenth century was not devoid of substance, not at all. Ironically its entelechy incorporated the dialectical movements that were to negate Christianity and eventually transubstantiate the latter's nature into the flesh and blood of modern science. But the drama remained incomprehensible to the protagonists and their critics.

Seventeenth-century poetry was more than hypertrophy of the figurative, then. It was sacramental, a tissue of symbols that endowed it with transformative efficacy. Galileo's *Sidereus nuncius* was the prosaic incarnation of the *concettismo* style. Galileo-Mercurius, the celestial messenger, was the sublime *concettista* who brought back to earth fantastic images of a universe whose fabric was at the same time material and imaginary. The images worked wonder in the mind of the reader, transforming the imagination into a telescope trainable on the vast expanses of cosmic consciousness. Originally sacrament means but mystery.[62] There was something mysterious that lay hidden in seventeenth-century poetry, a hieroglyph pertaining to divine things, which surpassed human reason. To understand its meaning one has to swallow its exacerbated figures, by hook or by crook. By signifying the divine mysteries sacramental poetry sanctifies the human that participates in its ritual. The sanctification of human reason performed by the poetic rituals of the seventeenth century has gone extinct in our culture. It has been brutalized by the mechanistic dogmas that were preached by the high priests of positivism, and which still hold sway in our educational institutions.

Galileo's 'Enigma' and Ours

Galileo was always concerned with the experience of poetry as a privileged locus for the fluid substance of individuality to emerge into consciousness. The sacramental force of poetry loomed large in his mind. His life-world brimmed with realities to be brought into existence. In his late years he wrote a poem staring at the abyss of death. In 1640, Antonio Malatesti, a Florentine jocular poet of some renown, published a book of poetry entitled *La Sfinge* [The Sphinx]. Galileo had the poems read to him (he was then totally blind).

62 Billot 1914, pp. 17–23.

Shortly thereafter he composed a sonnet entitled *Enimma* [Enigma], which he graciously sent to Malatesti.[63]

Galileo's sonnet (ca. 1640)[64]	Translation (by Dennis Looney)[65]
Enimma	*Enigma*
Mostro son io più strano e più difforme	Monster am I, stranger and more misshapen
Che l' arpia, la sirena o la chimera;	than the harpy, the siren, or the chimera.
Nè in terra, in aria, in aqua è alcuna fiera,	There is not a beast on land, in air or water
Ch' abbia di membra così varie forme.	whose limbs are of such varied forms.
Parte a parte non ho che sia conforme,	No part of me is the same size as any other part;
Più che s'una sia bianca e l' altra nera;	what's more: if one part is white, the other is black.
Spesso di cacciator dietro ho una schiera,	I often have a band of hunters behind me
Che de' miei piè van ritracciando l'orme.	who map out the traces of my tracks.
Nelle tenebre oscure è il mio soggiorno;	In the darkest gloom I take my rest,
Chè se dall' ombre al chiaro lume passo,	for if I pass from the shadows to bright light,
Tosto l' alma da me sen fugge, come	quickly the soul flees from me, just as
Sen fugge il sogno all' apparir del giorno;	the dream flees at the break of day.
E le mie membra disunite lasso,	And I exhaust my discombobulated limbs
E l' esser perdo, con la vita, e 'l nome.	and lose my essence, along with life and name.

Galileo's individuality is obfuscated by darkness, his rational soul chased by doubt. It is only when death occurs, the loss of personal identity (as hinted at by the last word of the poem), that the disarticulated members of Galileo's individuality will gain renewed unity and harmony in the light of after-death. The monster extended metaphor, with the imaginary hunters mapping out the traces of Galileo's tracks, captures the discombobulation, the jagged contour of the self's confrontation with the shadows of the unconscious. The self is fragmentary, made up of sundry pieces, its substance loose, impenetrable to reason. Rational knowledge in this life is dream. Only after biological

63 Cf. Favaro's comments on the episode, in Galilei 1890–1909, IX, p. 26.

64 Galilei 1890–1909, IX, p. 227.

65 Thanks to my colleague Dennis Looney for his splendid translation.

death, with the corruption of the material body, will the sublimation into the light of divine truth become final. In other words, the price human beings have to pay for truth is the loss of individuality, the undoing of the process of individuation, the liberation of the *anima* that ensouls the corruptible body. The sonnet's style is dominated by chromatic contrast, light/darkness, white/black, for the spectrum of colors that are perceived under normal circumstances had been obliterated by blindness, leaving the sense of sight floating in flickering darkness, bereft of the solidity of the surface color that is projected on surfaces and encapsulating the seer in the dimensionless depths of what Gestalt psychologists call 'the perception of film color'. The poem concludes with the intimation that the soul escapes from the self as a dream evanesces at dawn, and the voice, the saying-I, will lose both name and life, all of which does not seem to harmonize well with the Catholic dogma of the immortality of the created human soul. Indeed, David Wootton has called into doubt the orthodoxy of Galileo's Christian faith.[66] Orazio Ricasoli Rucellai reported that Galileo thought that light is the first beginning of nature since he believed that light was the ultimate expansion, that is, the ultimate rarefaction [of matter], the first principle from which all things would form (*tutte le cose si componessero…*) by condensation, reaching the densest and most impenetrable compression of the hardest stone.[67] Galileo's friend, Stefano della Bella, made a fascinating drawing, *Death carrying*

66 Cf. Wootton 2010, pp. 240–250.

67 Favaro 1907, p. 106. This is consistent with Galileo's theory of light in *Il saggiatore*. "E forse mentre l'assotigliamento e attrizione resta e si contiene dentro a i minimi quanti, il moto loro [i.e., the minimal particles of fire that Galileo calls *ignicoli*] è temporaneo, e la lor operazione calorifica solamente; che poi arrivando all' ultima ed altissima risoluzione in atomi realmente indivisibili, si crea la luce, di moto o vogliamo dire espansione e diffusione instantanea, e potente per la sua, non so s'io debba dire sottilità, rarità, immaterialità, o pure altra condizion diversa da tutte queste ed innominata, potente, dico, ad ingombrare spazii immensi" (Galilei 1890–1909, V, p. 352). The fact that Galileo thinks that material bodies are permeable to the *ignicoli* seems to imply that despite their apparent opaqueness they are permeable to light. But I think that what Galileo is really saying is that the ultimate constitution of material bodies is light itself, and that finite atoms or particles are conglomerations of infinitely small, non-quantifiable (*non quanti*), point-like corpuscles of light. The world of material bodies is made of light, Galileo intimates. This theory is similar to the doctrine we find in the *Pimandro*, the first treatise in *Corpus Hermeticum*, where man is said to be made of light and life. "Rettamente parli, perche Dio é vita et luce, et Padre, de'l quale é nato l'uomo. Se adunque tu comprenderai te medesimo esser composto di vita, et di luce, ancora trascenderai a la vita, et a la luce". Cf. Trimegisto 1549, p. 10.

a child (a theme he was fond of), which will help me introduce death in my meditation on Galileo's poem. But I appropriate the poetic image of Galileo's death from within my own life-world, and thus I picture it in my imagination overlaid with another intriguing image, a woodcut print by the German expressionist Ernst Barlach, who represents the same theme. In both cases death appears refracted through the prism of the dead child. Della Bella's death outrageously runs away with the saluting child, much the same way as consciousness departs from the discombobulated elderly individual, sinking into the abyss, or rising to the infinite summit of the divine essence. Barlach's dead child is really a corpse being buried in the maternal element. The seventeenth century looked at death as 'being carried away', while the twentieth-century artist observes it as the return, or perhaps the restitution of the dead child to the material/maternal element that receives the corpse, mother earth, as the common etymology of the Latin terms *materia* and *mater* suggests. Both images address a truth that escapes the power of rational thought. Both are essentially black and white images, though Della Bella's gestural drawing is partially covered in grey washes that convey some illusion of color. Galileo's "Chè se dall' ombre al chiaro lume passo", intimates a luminous passage (verbally hinted at with the verb *passo*) that cannot be described rationally. It is as if you wanted to remove the blackness from the pictures, reducing the chiaroscuro to pure light (*chiaro lume*), for then the volume of the figures in the pictures would evaporate, and the images themselves would collapse away from reality (which is in fact how Della Bella portrays the child, fading from perceptual reality and pale, as opposed to the sharply perceivable, fleshed, hence *real* death).

Fig. 1.2. Stefano della Bella (1610–1664). *Death carrying a child.* Medium: Pen and brown ink and brush and gray wash, over black chalk. Dimensions: 12 15/16 x 8 1/16in. (32.9 x 20.5cm). The Metropolitan Museum of Art, Credit Line: Gift of Peter Palumbo, 1983. Retrivable at: www.metmuseum.org

Fig. 1.3. Ernst Barlach (1870–1938). *Death of a Child (Kindertod)*. Date: 1919 Medium: Woodcut. Dimensions: block: 9 3/8 x 14 3/16 inches (23.8 x 36 cm). The Metropolitan Museum of Art, Credit Line: Bequest of Scofield Thayer, 1984. Retrivable at: www.metmuseum.org

On the verge of the abyss, I project myself unto Galileo's death, as I contemplate his passage from the vantage point of thinking forward to my own unimaginable death. I welcome the fleeting images popping up in my active imagination, such as the most poignant painting, *Departure of the adventurer* (Abfahrt des Abenteuerers [1939]), by the dying Paul Klee, a potent picture that sucks away my whole mind within its calm immensity, as Rucellai's testimonial now acquires an hitherto unsuspected meaning.[68] Galileo-Poimander teaches a cosmogony, singing in a voice that resonates with the echoes of Hermes Trismegistus. In the Italian edition of Marsilio Ficino's *Poimander* (Il Pimandro), translated by Tommaso Benci, a comment by Calcidius is reported on Hermes Trismegistus's death. "Until now, my friends, I have lived as a pilgrim, cast away from my country. When in a little while I will be free of the corporeal blemishes and take leave of you, do not cry as if I were dead, for

68 Augusto Alfani, the author of the only scholarly monograph on Rucellai to date, argued that this opinion is not to be found in Galileo's books, and that it was a hypothesis put forward by Galileo in oral settings and conversation (Alfani 1872, p. 142). A color reproduction of Klee's magical *Departure of the adventurer* is in Vry 2011, p. 81.

I will return to that blessed and great city, where all citizens will have to come because of the corruption of death. For, here is the highest prince, the only God, who makes his citizens blessed. Thus, this which most call life should rather be called death than life".[69] The Benci edition also published a magical Cabalistic text under the name of Raziel (*Testimonianza di Raziel*), describing the death of Trismegistus, and claiming that Hermes carried within himself the image of the father (who is also qualified as the abyss of piety), the abyss of the wisdom of the son, and the gift of the spirit of prophecy.[70] Hermes was the incarnation of the infinite abyss of being towards death. Wisdom and piety are the abyss that the claim of death presents the dying human being with, as their ultimate possibility of authentic being-in-the-world. Galileo's philosophical abjuration, "Parte a parte non ho che sia conforme, Più che s'una sia bianca e l' altra nera", evokes the problem of truth, but disbands its philosophical cohorts, once and for all, as the infantile dream of the possibility of *adequatio intellectus et rei.* No part conforms with any other part, not any more than the candor of white were to be adequate to speak the truth of blackness. Truth is the infinite abyss of wisdom and piety. But only the dying human becomes open to truth authentically. Its infinity is the revelation of the dream of philosophical truth.

The Discovering and Concealing Genius

Edmund Husserl spoke of Galileo as "at once a discovering and a concealing genius", who exposed the true mathematical structure of reality, while fatefully though unconsciously covering up the untruth of the immediate presence of the life-world.[71] He also stressed that the crisis of European sciences brought

69 "Io, sino a qui, o figliuoli, sono vissuto peregrine; et sbandito dalla patria. Et quando di qui a poco sciolto da le macule del corpo, da voi mi partiró, vedrete di non mi piangere, come morto: imperó che io ritorno a quella ottima, Et beata cittá: a la quale debbono venire tutti i cittadini, per la corruzione della morte. Imperó che quivi è sommo principe, solo DIO: il quale i suoi cittadini di maravigliosa suavitá riempie. Adunche questa, la quale molti chiamano vita; piú tosto si debbe dire morte che vita" (Trimegisto 1549, not paginated, see section entitled "Calcidio nella ultima parte del suo secondo libro dice di Mercurio Trimegisto cosí").

70 "Trismegisto [...] ebbe in se la imagine del padre, lo abisso della sapienza del figliuolo, e il dono dello spirito della profezia. [...] Visse pia et religiosamente; et piisima et religiosisimamente morì" (Trimegisto 1549, not paginated, see section entitled "*Testimonianza di Raziel*"). Cf. Avilès 1997, on the magical-Cabalistic background of the *Liber Razielis*.

71 Husserl 1970, p. 52.

about by positivism consisted in the loss of meaning of the sciences for human existence. In conclusion, I would like to suggest that Husserl's concern with an alleged crisis of the European sciences around the 1930s is still very much meaningful for us today, more than eighty years later. Husserl compared science to a machine that anyone can learn how to operate correctly "without in the least understanding the inner possibility and necessity of this sort of accomplishment".[72] Though the Husserlian notion of a *European* science is no longer applicable to industrialized societies and global economies, the monster machine that was built by the nineteenth century and transformed by European positivism into a Galilean myth of origin has worked to obscure what I will call the 'numinosity' of hermeticism. It is a loss whose consequences we still suffer today. It is imperative to reintegrate that numinosity in our collective consciousness if we are to revive the deeper meanings, the inner possibility and necessity, of Galileo's project for human existence.

72 Husserl 1970, p. 52.

· 2 ·

HERMES

In this chapter I address being-in-the-world of *hereticality*, being-in-the-world of *libertinism*, and being-in-the-world of *hermeticism* as three modalities of projecting possibilities of understanding in early modern life-worlds. These three modalities drew my attention to one structural element of being-in-the-world that should be considered in relation to the history and philosophy of early modern science. It is the element of doing violence, here intended, in a first approximation, as the action of constraining or forcing unnatural or undesired change upon something or somebody by physical or cognitive means. It can be performed by both epistemic and physical coercion, and its uncanny underlying the possibilities of understanding requires investigating its historical modalities. Is Western philosophy in denial about its history of violence? Projecting possibilities of understanding does not mean grasping facts of experience, or objects or ideas in the sense of explaining them, causally, mathematically, or otherwise. Rather I would like to explore understanding that implicates being-in-the-world in its ownmost possibility for being authentically, that is, for renouncing the lynching of the resisting other by enforcing its logic of closure, and to examine how violence and the possibility of acquiescence to it, or its rejection play into this more radical possibility for authentic existence. It is the drama of the obligation towards both one's authentic self and one's submissive integration into alterity.

Hereticality

To undertake this meditation on philosophy's denial of violence in early modern life-worlds, I begin with a reflection on violence in my own life-world, in order to break through the barrier of silence that philosophy has erected. Especially because the term *heresy* has become a weasel word in modern parlance, I must emphasize from the outset that I will restrict my inquiry to early modern Italian *hereticality*, by which I refer to the phenomenon demarcated and investigated by the Italian historian and philosopher, Delio Cantimori, in his epoch-making monograph, published in 1939, on the Italian heretics of the sixteenth century (as well as in a number of minor studies).[1] I must begin with heresy, firstly because it brings up the question of violence most clearly, but secondly and most importantly because Cantimori was the first and foremost Italian intellectual (a philosopher turned historian) who consciously intertwined his work on the heretics with a hermeneutics involving the question of violence in his time and, crucially, the question of his own Fascist militancy in the context of the violence that had dominated Italian social-political history before and after Mussolini seized power. This 'secret' consciousness, however, can only be reconstructed, or divined a posteriori, as its efficacy was disguised by Cantimori in his celebrated book. Although the history of the Italian heretics of the sixteenth century has been enriched over the last few decades, Cantimori's work remains absolutely fundamental because the enrichment has, regrettably, in my view, proceeded merely by way of unearthing empirical, namely, archival material, in the framework of a historiography informed by the positivist construal of history as a natural science whose criterion of objectivity is the philological restoration of archival material. What is missing in these more recent efforts is the philosophical scrutiny and the hermeneutic consciousness which had sustained Cantimori's trail-blazing research. Yet Michele Ciliberto, a historian of Renaissance philosophy, observed that Cantimori was not only a foremost historian but a philosopher of history who entered the political arena while being perfectly cognizant that historiography and political praxis are the two polarities of an eternal tension.[2] It is no wonder that the question of violence

1 Cantimori 1992 (1st ed. 1939). This new edition of Cantimori's book contains excellent introductory material by Adriano Prosperi, discussing both Cantimori as a historian and a philosopher and the tortuous history of Cantimori's monograph. The volume is accompanied by a useful bibliography that offers guidance to the field of Italian hereticality.

2 "Cantimori non è stato solo un grande studioso di storia, e uno dei pochi, moderni filosofi della storia che la cultura Italiana abbia avuto nel Novecento; è stato anche un

and of its philosophical implications has gone unnoticed among the positivist historians who followed in Cantimori's footsteps after the second world-war.

In 1978, two years after Ciliberto published his foreboding note on Cantimori, highlighting the philosopher of Italian hereticality, Italy entered the gloomiest weeks of its entire post-war political history. The Christian Democrat leader, Aldo Moro, was kidnapped by the Red Brigades, a group of left-wing, self-styled revolutionaries, but labeled 'terrorists' by the Italian government. At the time I was a teen-ager student in the "Liceo Classico" (humanities high school). As the news of Moro's kidnapping broke, like a thunderbolt, classrooms emptied, and we invaded the commons of the school as if an immediate urgency about a fact of which we knew nothing called upon us to convene hurriedly. We held a general meeting to discuss what we clearly perceived as an apocalyptic situation. A scant two months later Moro's corpse was discovered in the trunk of an abandoned car in Rome's historic center. For me the image of a hurricane of violence breaking into political history is consigned forever to the memories of those days, the horror of the massacre of the security agents escorting Moro in the course of the Red Brigades's attack, and above all to a historic picture, almost like an emblem of the nature of violence, that the Red Brigades had relinquished of the prisoner sitting in front of a banner displaying their symbol. The question I wish to raise in retrospect is whether the Red Brigades can be considered as a twentieth-century heretical movement, and if so whether the modality of being-in-the-world of twentieth-century political-ideological heresy, namely, that which we label 'terrorism', projects possibilities of understanding, affirming the eternal tension of historiography and political action. More broadly, is violence a manifestation of, or response to, philosophical heresy? To reiterate, I raise the question of philosophy's denial of the possibility of its being a form of violence in the life-world. Thus, is violence, both physical and epistemic, underlying modalities of being-in-the-world that projects possibilities of understanding? The question is not preposterous, for I need to face the reality of my perception of violence, within my own life-world, if I am to encounter meaningfully the perceived violence of hereticality in early

intellettuale che ha cercato di stabilire, contemporaneamente, fin dalle origini, nell' epoca dei grandi movimenti di massa, un rapporto di tipo nuovo, positivo, non mediato, con la politica e le sue organizzazioni. Cantimori è stato costantemente consapevole delle novità indotte dalla politica sul terreno della ricerca storica, della cultura e della sua 'autonomia'. [...] politica e storiografia, teoria e politica *in actu* sono poli essenziali di una tensione permanente" (Ciliberto 1976, p. 57).

modern life-worlds. It might be objected that I should strive to be free of prejudices and thus do my best to forget my perception of reality as it appears to me within my life-world. But to this objection I would like to answer, following Benedetto Croce, that within my being-in-the-world my convictions and perceptions cannot have the character of prejudices. Instead they have the force of liberating my judgment from prejudices, of freeing my mind for my ownmost possibilities of understanding, yet a mind that is not Edenic innocence but the product of laborious reflection.[3]

Fig. 2.1. [Lynching, Russellville, Kentucky] Artist: Minor B. Wade (American, 1874–1932). Date: 1908. Medium: Gelatin silver print. Dimensions: Image: 11.8 x 9.1 cm (4 5/8 x 3 9/16 in.) Credit Line: Gilman Collection, Gift of The Howard Gilman Foundation, 2005. Retrievable at: http://www.metmuseum.org

3 "Ma io desidererei che si considerasse che le mie 'convinzioni' non possono avere, nel mio animo, il carattere di pregiudizi, ma per l'appunto di quella liberazione, che si richiede, dai pregiudizi, di quella mente sgombra e pura, necessaria per intendere i fatti storici, e che non è un' innocenza edenica, come taluni immaginano, ma una faticosa formazione di cultura" (Croce 1912, p. 19).

As an intellectual exile living in the United States I became dramatically acquainted with another modality of being-in-the-world, which I have integrated into my life-world and thus will affect my appropriation of the modalities of being-in-the-world of sixteenth-century Italian hereticality. I am referring to racial terrorism. There is a picture on the website of the Metropolitan Museum of Art entitled "Lynching, Russellville, Kentucky". The caption informs us that "[t]his photograph is brutal testament to racial terrorism in America". The facts of the case are drawn from a small article that appeared in the "New York Times" on August 2, 1908, the same day the photograph was made by a local journalist. On the previous night, one hundred white men had entered the Russellville, Kentucky, jail and demanded that four black sharecroppers who had been detained for "disturbing the peace" be turned over to them. The men were accused by the mob of *expressing sympathy* for a fellow sharecropper who, in self-defense, had killed the white farmer for whom he worked. The jailer complied, and Virgil, Robert, and Thomas Jones and Joseph Riley were taken to a cedar tree and summarily lynched. The text of the note pinned to one of the bodies was also inscribed on the verso of the photograph: "Let this be a warning to you niggers to let white people alone or you will go the same way".[4] The practice of racial terrorism persisted until at least the early 1940s. The question I wish to raise in this regard is whether the African-American victims of lynching terror were perceived as heretical in relation to a society in which white supremacy was a political, theological, and scientific dogma. Thus, what are the modalities of being-in-the-world of an African-American victim of racial terror? What possibilities of understanding are projected such that are perceived as a destabilizing threat to white supremacy? My reflection on violence as structural element of being-in-the-world raises the question of the meaning of the word 'violence' as linguistic instrumentation that being-in-the-world finds at its disposal in its character of everyday anonymity. It is the humanistic problem of *verbum* (word) and *res* (thing) that takes priority over the problem of the essence of things. What is the *res* that corresponds to the word 'violence'? Is there such thing as an essence corresponding to a *verbum* that vouchsafes its communicability? It appears that the word 'violence' in its everyday usage in multiethnic social contexts does not have any stable meaning that transcends history.

4 See the caption of *[Lynching, Russellville, Kentucky]* Artist: Minor B. Wade (American, 1874–1932). Date: 1908. Medium: Gelatin silver print. Dimensions: Image: 11.8 x 9.1 cm (4 5/8 x 3 9/16 in.) Credit Line: Gilman Collection, Gift of The Howard Gilman Foundation, 2005. Retrievable at: http://www.metmuseum.org. The emphasis is mine.

Its protean nature is a manifestation of the ethnicity of reason. To the practitioners of racial terrorism against African-American people, the perceived threat posed by the latter, physical or verbal or epistemic ('...expressing sympathy'), challenged the orthodoxy of white supremacy. Understanding one's being-in-the-world as possibility of 'expressing sympathy' means a decision to live authentically towards one's death. To the Italian government and to the vast majority of the Italian population at the time, including myself, the violence of the Red Brigades appeared as posing a threat to the fabric of Italian society and to its democratic ideals. Four decades later I have not changed my mind, though I find myself more and more disempowered by the brutality of the violence that 'democratic' societies practice towards diversity. Moreover, subsequent history has proved that the political class targeted by the Italian 'terrorists' was itself the embodiment of a dogmatic, oppressive, and shamefully corrupt oligarchy. Hence I push myself to ask the question: Might the Red Brigades be seen, in hindsight, as heretics embracing what we perceived at the time as unspeakable violence, but which they saw as a means of overturning what they perceived as an imperialist and terrorist political class? Also the weasel word 'heretic', then, does not seem to have an identity carrying a communicable essence. It manifests its character of everyday anonymity. What is the *res* of heresy?

Cantimori demarcated the phenomenon of Italian hereticality very strictly.[5] But it was a problematic choice that ultimately had to do not so much with the archives as with the philosophy of history and the role of violence in history. More precisely, it had to do with Cantimori's subject matter, which was squarely the history and philosophy of Italian heretical thought and, more profoundly, albeit not consciously articulated, the role of violence in thought and science, though a science that the twentieth century disowned. It was the queen of early modern sciences, Christian theology, as it was being reconfigured in the age of the emergence of Christian denominations. It was the controversialist theology fuelling the European religious

5 This question of demarcation has an interesting history per se which is intimated in the title of the book. In 1939, Cantimori defined the Italian heretics as "ribelli ad ogni forma di comunione ecclesiastica" (Cantimori 1992 (1st ed. 1939), p. 5), emphasizing their self-conscious rejection of membership in organized churches, but in the preface to the German translation of the book, in 1949, he indicated that the criterion for labeling them as heretic was "il giudizio delle Chiese o degli uomini dirigenti delle Chiese, cattolica e acattoliche, contemporanee" (*ibid.* p. 9), this time emphasizing an external criterion based on the judgment of their contemporaries.

wars and religiously motivated genocide.[6] I surmise that for a period of time Cantimori saw in the fledgling Fascist movement the political and philosophical renewal that was to revolutionize the history of Europe. But the book was written speedily, at the outbreak of second world war, when already his faith in Fascism was vacillating, and in all probability he had transformed himself into an apostatical Fascist. Cantimori, the crypto-heretic, was born. He would soon join the Italian Communist Party. You can write powerfully only about what you are experiencing deep down as innermost possibility of being-in-the-world. And Cantimori's *Eretici* is a powerful book, the enduring masterpiece of Italian philosophically-conscious historiography in the twentieth century. It cannot be an accident, therefore, that such a book was written when its author withdrew from inauthentic existence, turning towards hereticality as a more authentic mode of projecting understanding within the dramatically changing life-world of the late 1930s.

From the outset Cantimori underscores the crucial importance of two places of origination of Italian hereticality. Its roots are to be found in both the humanistic culture that opposed late medieval scholasticism and the Neoplatonic revival promoted by the Florentine circle of Marsilio Ficino and Pico della Mirandola. The humanistic interest in theology and its bearing on ethics was especially marked in Lorenzo Valla, who aptly deployed his philological scrutiny, aimed at precisely defining the meaning of words, not in view of stylistic concerns but because he was motivated by a deeper anxiety about the relationship between human being and God, and about the social and political intercourse among fellow human beings. In other words, Valla drew his philological inspiration from the jurisprudential and theological spheres.[7] Cantimori calls attention to Valla's attack on Boethius, notably the latter's conception of persona. In this regard I find Valla's vehement yet lucid polemic exceptionally suited for observing the mechanics of violence in humanistic argument. "Persona, according to Boethius," Valla claims, "is an individual

---

6 Cantimori gave a dramatic and courageous testimony, in retrospect, of his own political involvement with Fascism, in regard to his cold attitude towards, and fascination with, the violence that Fascist groups of militants (*squadristi*) perpetrated in the course of punitive actions against socialist and communist targets (though he does not say that he ever took part in these expeditions). See Cantimori 1992 (1st ed. 1939), p. xxii.

7 "Ed è caratteristico come il Valla procedesse non sulla base dell' eloquenza, come erroneamente si suol dire seguendo uno schema tradizionale, ma sulla base del diritto e della teologia: che impegnavano i valori findamentali della vita, i rapporti fra uomo e uomo e uomo e Dio". Cantimori 1992 (1st ed. 1939), p. 16.

substance of immutable nature".[8] According to Valla, Boethius was unable to locate persona in any of the species in the subdivision of accident, and therefore categorized it as belonging in the genus substance. But, he charges, "to this Roman, I will show that he does not know how to speak Roman".[9] He states that he cannot see why persona should not be called a quality regardless of whether we speak of human beings or of God. In fact in everyday parlance when we speak of persona in reference to human beings, we mean the quality by which one individual differs from another, which can be in the soul or in the body or in external things such as wealth or poverty. We speak of persona when we refer to the characters that actors play on the stage. However, it is not the case, Valla insists, that to each single individual a single persona corresponds, for multiple personas will always be found in an individual. I am human, liberal, timid, genial, stupid, sleepy, learned, old, father in law, brother in law, father, son, brother, beautiful, strong, wealthy, noble, and occasionally the opposite of all of these. In fact, in relation to the wealthiest person, for example, I will not be wealthy, and therefore it is obvious that persona is not absolute but relative. To myself, then, multiple and different personas, or qualities adhere, but only one substance. I attribute to God three qualities, says Valla, and these I claim to be persona, despite the fact that Boethius wishes to shut me up by urging that quality is that attribute which can be separated from its subject without the latter's corruption.[10] But Valla points out that in Latin everyday parlance, following the consensus of all authorities, we speak of the light and heat and scintillation (*vibratio*) of the sun as qualities in the sun. Qualities in this sense are those in God that constitute his persona (not his substance), and which cannot be separated from him, not any more than light and heat and scintillation can be separated from the sun. The fact of the matter, concludes Valla, is that Boethius taught us how to speak barbarian.[11] How does Valla's being-in-the-world project possibilities of understanding? The latter does not mean grasping facts of philology only, based on the consensus of the Latin authorities, or grasping philosophical and theological ideas in the sense of explaining them philologically. Understanding the correct way of speaking Latin implicates Valla's being-in-the-world in its ownmost possibility for being authentically, for in his case the speaking of correct theological

8 "Persona est (inquit Boethius) incommutabilis natura, individua substantia". Valla 1543, p. 215.

9 Valla 1543, p. 215.

10 Valla 1543, pp. 215–216.

11 Valla 1543, p. 216.

words is not embellishment, or the art of oratory for its own sake, instead it is the authentic possibility of being Valla insofar as authentic existence resides for him in these possibilities of correct speaking. Yet Valla appears aggressive in his assault on Boethius. He pours scorn on the Roman by teaching him how to speak Roman. His ground for launching this epistemic assault is the analogy with the sun's qualities. The transfer of semantic appropriateness from the domain of the sun to the domain of God is a dramatic passage from natural philosophy to theology. It is a possibility of understanding by imaginatively translating being in the sun to being in God. Valla's imagination casts his own shadow upon the human landscape, which is the violence perpetrated by Valla that Valla himself cannot see. For, how does Valla speak Latin *correctly*? By gathering the appropriate meaning of words from listening to the chorus of authorities. The choral dimension of authority is the out-there in the world where Valla is magnetically attracted to and where he loses himself. He is dragged away by the anonymity of the chorus singing in unison. The Latin word *persona* resonates with multiple meanings, as its ambiguous etymology testifies; for example, in the scenic context, it refers to the mask used by the Roman actors, which covered their head, and might be different according to different characters. Valla himself intimates that our being multiple personas implies our possibly being thrown out of our ownmost possibility of authentic existence. In fact he hastens to add that we are multiple personas but one substance. He stitches this substance to our being hurriedly (note that his chapter on persona ended up on the Index of Forbidden Books for Catholics).[12] He has come close to the abyss. He sees the infinite emptiness of the abyss of persona, the multiple resonance of the infinite voices we speak inauthentically, or rather the undifferentiated multitude of voices that sing in unison through the aperture of our own mouth. We cover our head with masks through which our voice resonates, reverberating a cacophony, and we thus lose ourselves. We exist in the inauthenticity of the others that blots out the possibility of our sighting the violence, physical and epistemic, that we blindly perpetrate on both ourselves and the others.

The second root of Italian hereticality was the fifteenth-century Neoplatonism of Marsilio Ficino.[13] According to Cantimori, it was Ficino's attempt to demonstrate the possibility of reconciling Christian religion with Neoplatonism that catalyzed a process of consciousness, leading to the emergence of Italian hereticality, even though initially that consciousness took a form

12 Cantimori 1992 (1st ed. 1939), p. 240.

13 Cantimori 1936.

subservient to scholastic philosophy, that is, it accepted Christian dogma as its criterion of truth. However, Ficino was able to mitigate this epistemic requirement, in that he realized that it was no longer sufficient or desirable to demonstrate the accord between a specific philosophical doctrine and a specific Christian dogma. Impiety was rampant, and, according to Ficino, what was really needed, short of a miracle, was a structural framework for reconciling the whole of Christianity, the superior religion, with the whole corpus of Neoplatonic doctrines.[14] The ideas of the Italian humanist Celio Secondo Curione, professor of humanities at the University of Basel, offer an exquisite example of how an exponent of Italian hereticality positioned himself alongside the Neoplatonism delineated by Cantimori. Curione's inspiration is generally pantheistic and anti-Trinitarian, though this tendency was best expressed not so much in his polemical writings or biblical exegesis as in the form of a humanistic fable entitled "*Araneus, seu de providentia dei*" (Spider, on divine providence).[15] It is a work that deserves serious philosophical scrutiny though I will only be able to touch on a few relevant points. It showcases a flamboyant surrealism that mingles and mangles, in a florid prose, almost like a display of Ciceronian fireworks, Neoplatonic philosophy, Pythagorean numerology, natural history, mythology, theology, and scriptural exegesis. The propagandistic subtitle highlights that it is a little jewel (*libellus vere aureus*). No doubt it is. I will compare and contrast Curione's little known heretical work of art with a more celebrated, poorly understood, much shorter, but equally avant-garde, equally heretical, and yet ever more mysterious jewel fable narrated by Galileo in *Il saggiatore*. I emphasize that I do not intend to suggest that Galileo knew about Curione's work, either directly or indirectly, or that he was consciously placing himself in that heretical tradition. I am not looking for sources, and not because I think they are worthless when they can be established (quite the contrary), but because the hermeneutic of encounter strives to be sensitive to all possibilities of understanding that are projected as modalities of being-in-the-world. As such it is always only 'subjectively' sensitive to the sources, and this 'subjectivity', far from being a hindrance to encounter, is the light that is cast upon the life-world by authentic existence. Thus, the notion that Curione's fable might be a source of ideas for Galileo's, and that, if it is not a source, then it is preposterous to put it forward, does not stand up to scrutiny either way. No source can ever explain or forbid thought. The belief in the explanatory power of the source is the residue of the positivist

14 Cantimori 1936, pp. 545–546.

15 Curione 1544, pp. 1–88. Cf. Cantimori 1992 (1st ed. 1939), p. 108–112.

dogma that an alleged precedent, which can be hypothesized on the basis of textual similarity or other evidence, constitutes empirical evidence on which the objective history of thought must be founded as positive science. Benedetto Croce warned his positivist critics, who accused him of neglecting or distorting the sources of Giambattista Vico's philosophy, that "the search for precedents can never explain the new thought which was erected on them; let alone do the precedents diminish its value. That sort of information, though on the one hand it enriches our knowledge of the history of philosophy, on the other, in regard to the specific thought that is being investigated, is devoid of efficacy. Useful for writing the philosopher's biography, it is useless in order to understand the meaning of the new theory, which must be deduced from the new problem that that theory tackled and aimed to solve".[16] Unabashed I venture to live-through the process of de-familiarization I find myself thrown into. I am a prey entangled in the web of Curione's *Araneus* and ready to be digested by the spider. Here is an excerpt from his narrative.

> There once lived, in a very solitary place, a man endowed by nature with a very perspicuous ingenuity and an extraordinary curiosity. He raised many birds for his own pleasure, much enjoying their songs, and he marveled at the happy contrivance by which they would transform at will the air they breathed into a variety of very sweet songs. It happened one night that close to his house he chanced to hear a delicate sound, and, being unable to imagine what it could be except some small bird, he set out to capture it. Arriving at the road, he found a shepherd boy who was blowing into a kind of hollow stick and moving his fingers about on the wood, thus drawing from it a variety of sounds similar to those of a bird, though by quite a different method. Astonished, and led on by his natural curiosity, he gave the boy a calf in exchange for his whistle and retired to meditate. Realizing that if he had not chanced to meet the boy he would never have learned of the existence of two methods for forming musical notes and very sweet songs, he tried traveling far from his home in the hope of meeting with some new adventure. The very next day he happened to pass near a small hut, and, hearing a similar tone within, he went inside to find out whether it was a whistle or a blackbird. There he found a boy holding a bow in his right hand and sawing upon some gut strings stretched upon a concave piece of wood. The fingers of the left hand (which supported the instrument) were moving, and without blowing the boy was drawing from it various notes, and most sweet ones too. Now, you who are participating in this man's intelligence and sharing in his curiosity, judge his

16 Croce 1912, p. 11. Croce's concession in regard to the philosopher's biography was a scholarly restrained way of dismissing the superstitious faith of the positivists in the virtues of history as science. It is paradoxical that his cautionary warning has gone virtually unheeded in the Italian (and especially Galilean oriented) historiography of science for more than a century.

> astonishment! Finding himself to have two unexpected new ways of forming tones and melodies, he began to believe that others might be found in nature. [...] Again, impelled by curiosity, he entered an inn expecting to see someone lightly bowing the strings of a violin, and instead saw a man rubbing the tip of his finger round the rim of a goblet and drawing forth from it a very sweet sound. And later he observed that wasps, mosquitoes, and flies did not form separate notes from their breaths, as did his original birds, but made steady tones by the swift beating of their wings. In proportion as his puzzlement grew, his belief weakened that he knew how sounds were generated; nor could all his previous experience have sufficed to make him understand or even believe that crickets, which do not fly, could draw their sweet and sonorous shrilling not from breath but from a scraping of wings. And when [...] he believed that he had seen everything, he found himself more than ever wrapped in ignorance and bafflement upon capturing in his hand a cicada, for neither by closing its mouth nor by stopping its wings could he diminish its intensely strident sound, and yet he could not see it move either its scales or any other parts. At length, lifting up the armor of its chest and seeing beneath it some thin, hard ligaments, thinking that the sound came from them, he reduced himself to breaking them in order to silence it. But it was in vain until, driving the needle deeper, he pierced the creature and took away its life with its voice, so that even then he could not make sure whether the song had originated in those ligaments. Thereupon he was reduced to such diffidence on his knowledge that when asked how sounds are generated he used to reply, in deference, that although he knew some of the ways, he was certain that many more existed which were unknown and unimaginable.[17]

Piccolino and Wade suggest that the fable conveys an optimistic meaning.[18]

For them the "final episode with the cicada raises a series of questions and leaves the reader somewhat puzzled: the insect dies after 'the amour of its chest' is pierced in an attempt to discover the mechanism of sound production. This alludes to the potentially destructive nature of scientific experiments. [...] From a playful activity, the scientific enterprise changes its character becoming a complex investigation, with uncertain results, needing an active and potentially destructive intervention, in order to broaden the vision of

17 Piccolino & Wade 2014, pp. 5–7. I have modified this elegant translation in a number of places. Cf. the original in Galilei 1890–1909, VI, pp. 279–281.

18 "As happens on all voyages, the path to knowledge is open to adventures and unexpected discoveries, sometimes beyond the limits of established categories and predefined mental schemes. It can thus induce wonder and amazement, but nevertheless it might not prevent us from discovery what we are looking for in a simple and direct way. [...] On the other hand, those who do not feel the need to begin their voyage of knowledge will try vainly to subjugate the variety of natural phenomena to their a priori schemes". Piccolino & Wade 2014, pp. 8–9.

reality".[19] Mario Biagioli, thought-provokingly, dubbed the man of the fable a 'philosophical brute'.[20] Science is brutality in its destructive intervention upon the world. Crystal Hall believes that the fable indicates that the "truth is discovered" rather than constructed.[21] All these meanings lived in the consciousness of the interpreters as, I believe, they approached the fable from a micro-historical perspective focused on Galileo's text and its immediate social and literary surroundings. I would add that there seems to be an awkward disavowal both of empirical knowledge ("nor could all his previous experience have sufficed to make him understand…") and of the possibilities of knowing nature ("…many more existed which were unknown and *unimaginable*"). Above all the fable's denouement points to a failure of inquiry terminating in the killing. The mystery of the fable urges that we seek an immersive experience of de-familiarization while recognizing that *nature* has not revealed her secret to the brutal philosopher doing her violence, further that we do not know what the *res* really is of the *verbum* 'nature' employed by Galileo, and that she knows (possibly infinitely) many mechanisms, some of which unimaginable, for producing sound ("…he began to believe that others might be found in *nature*"). The word 'nature' was common, according to the 1623 edition of the *Vocabolario della Crusca* (the same year as *Il saggiatore*). The entry notes that it embraces all forms of things, both essences and causes, and reports a quotation from Boccaccio, distinguishing *natura naturante* (God), *natura generante* (the heavens influencing the sublunary region), and *natura naturata* (the created world).[22] In the *Dialogo* (1632), Galileo's usage of *natura* (when it does not refer to the *nature of something*) is almost invariably coupled with the word 'God'. They come together, almost like king and queen, an alchemical hierogamy, as if playing the same creative and ordering role in the universe.[23] Their identity is intimated. I believe that Antonio Rocco was probably correct when he portrayed Galileo as a de facto pantheist. Rocco attacked Galileo's 'Platonic myth' (recounted in the *Dialogo*) regarding the origin of the current order of the celestial bodies in the cosmos, in which

19 Piccolino & Wade 2014, p. 10.

20 Biagioli 1993, p. 304.

21 Hall 2013, p. 127.

22 Vocabolario della Crusca 1623, p. 538.

23 Galilei 1890–1909, VII. For example, "…che alla natura e a Dio fusse impossibile il conferir quella velocità" (*ibid.*, p. 45), "…che così mi pare che ricerchi la ricchezza della natura e l'onnipotenza del Creatore e Governatore" (*ibid.*, p. 126), "…ma ci contentassimo di esser certi che Iddio e la natura talmente si occupa al governo delle cose umane" (*ibid.*, p. 394).

Galileo had narrated a cosmogony based on the free-fall acceleration of the celestial bodies. It is worth reading Rocco's allusive passage. "Is it not more reasonable to say that since God himself is from eternity, nature itself is from eternity? And that she is but God's instrument, immutable because of His immutability, predisposed by His wisdom? And perhaps God and nature are different only in their name, with an accidental difference in their effects? That is, while we say 'God' we intend that supreme entity, first, independent, unique in itself, infinite, optimal, most happy, so that He himself is nature with the instruments of the second causes that he deploys at will. And if this is the case, why should we not say that as nature operated for eternity, so she also operates at present, and in the same way she does at present, she has been doing forever?"[24] I propose that, if we enlarge our perspective, expanding Galileo's familiar surroundings or dwelling-place to include the undercurrents of Italian hereticality, Curione's fable, *Araneus*, will serve the purpose of estrangement, freeing the possibility of reactivating the meaning of nature's works in the cicada fable.

The miniature cosmos of insects de-familiarizes being-in-the-world by calling upon it to alienate its most cherished possessions, the ones which constitute its at-homeness. The contemplation of an insect's mode of life, the artful accruements by which insects weave the fabric of their existence, undergoes a withdrawal of world as it suspends the familiarity by which being-in-the-world experiences its own sense of belonging in nature. Curione notes that we call these creatures 'insects' because of their indentations and fissures, which challenge our expectation of perception and pose the question of the *ratio* (the reason, the plan, the order) which nature follows in creating them.[25] Indeed, he emphasizes the most spectacular feature of the ratio of insects, which is their savage, harsh voice.[26] It is a voice that is heard as extraneous to

24 "Non è egli più ragionevole il dire, che essendo l'istesso Iddio che fu ab eterno, sia anco l'istessa natura che fu? e che ella altro non sia che instromento dell'istesso Dio, immutabile dalla Sua immutabilità, ordinato dalla Sua sapienza? e forse Iddio e la natura differiscono solo di nome, con accidental diversità negli effetti? cioè, che dicendo Iddio, s'intenda quella entità suprema, prima, independente, unica in sè stessa, infinita, ottima, felicissima, e natura sia egli medesimo, con gli stromenti delle cagioni seconde che a Suo voler impiega? E se ciò è vero, perchè conseguentemente non diciamo, che come ab eterno operò la natura, così operi anco a i tempi nostri? e come a i tempi nostri così facesse ab eterno?" Cf. Galilei 1890–1909, VII, p. 601.

25 Curione 1544, p. 22.

26 "Denique quanta arte, quibusve instrumentis, aut quonam loco truculentam illam in nonnullis, et pro rata portione maximam vocem, atque sonum ingeneravit?" (Curione 1544, pp. 12–13).

the human world, a mysterious call from an orderly though unfamiliar arrangement of things. The proportion of the lengths of the anterior to the posterior legs of certain spiders is the sesquitertia proportion. According to Curione, the sesquitertia is a special proportion. In it enter the monad, the dyad, the triad, and the tetrad, which add up to form the number ten. The monad is the origin and cause of all numbers, converging into the constitution of all numbers to perfect their nature, being itself instituted by the most perfect nature, which we call God, and given to us as the symbol of piety. The dyad, created by the perfect one first, signifies the sensory world in which multiplicity began to manifest itself to us. Since generation cannot happen without the copulation of two, it is likely that the dyad is a figure of generation. In God everything is one, in the world everything is two, as the perlustrating mind will easily conceive, says Curione.[27] But spiders are not only immensely capable in their everydayness, their existence shows the possession of an uncanny 'sensus futuri', a sense of the presence of the future, which allows them to cope with events that threat their existence, such as an approaching thunderstorm. What is it that makes the future present to them? It is that divine force that is diffused throughout creation, the mind penetrating the body. But certainly this divine operation remains mysterious, we do not know the way in which God performs this activity. Yet it is all the more admirable that divine operation is equally present in some of the minutest creatures. And why is that? Because God makes himself visible through these disorienting modalities of being. It is a possibility of apprehending God through his de-familiarizing vestiges that humans can seize upon, since God cannot become manifest to us directly, in our own everydayness, like the sun, which cannot be observed by the naked eye. God in his wholeness became Christ so as to appear to us in the familiar guise of *persona*.[28] But God is always 'visible' to us through the mysterious operations that insects such as spiders perform. Curione's pantheistic tendency is evident. "He will not be mistaken who realizes that that which is, that which coheres, and which consists in itself, is God".[29] Nature tenders us the possibility of contemplating God, calling us to respond to the call proffered by him from the outside of what is most familiar to us, the miniature cosmos of insects, because nature manifests God in his externality.

27 Curione 1544, pp. 18–19.

28 Curione 1544, pp. 25–31.

29 "...non aberrabit opinor, qui quod est, quo cohaeret, et constat, Deum esse fatebitur" (Curione 1544, p. 58).

The mystery of the cicada's harsh voice is enveloped in the obscure origin of the sound calling from the outside of the familiar human world. That mystery the investigator of nature fails to solve. The perpetrator of the violence that kills the cicada has stepped outside of the local order of divine providence. He defiles nature by killing. The harsh voice of the cicada is the voice of God calling upon humans from the inhospitable desert in which humans cannot dwell (truly *vox clamantis in deserto*). The cicada fable is recounted by Galileo in the *Saggiatore* in the context of the polemics on the origin of comets. It is an interlude in which the violence of argument and counterargument is suspended temporarily. The heretic Curione elevated the humanistic fable to the rank of a theological microscope that opens a vista on the natural vestiges of providence. The destructive force of epistemic violence constantly rupturing through the arguments of the *Saggiatore*, which subjugates Galileo's ego, is subdued only in the fable. Violent being-in-the-world casts a shadow on the possibility of understanding, not because it shuts down cognition as the grasping of the world, which it certainly does, but primarily because it prevents the de-familiarization process that brings up the world in plain view for scrutiny. Imagination retains the originary force of projection that liberates the ego from the constriction of epistemic violence. In the imagination hereticality is recovered as an aboriginal mode of being-in-the-world that is non violent. I portend the future and the past in the imagination. I bring time under the purview of the present, as the spider does, mysteriously, in my aboriginal being-in-the-world of nature that expresses sympathy rather than curiosity. The violence of the curious investigator who kills the cicada because his ego is tormented by his failing to grasp the mechanism of God's voice, is as abhorrent as the violence of the racial terrorists who silence the expression of sympathy. Our ego is the endogenous white supremacist that kills our ownmost possibility of authentic existence.

As a non-coercive reflection on the theme of violence and hereticality I would like to offer the following dream-fable. Since I have taken a phenomenological stance throughout this research I must respect the totality of experience that calls upon me, without prejudice about what should be accepted as relevant to my understanding and what should be disregarded.[30]

30 This dream came to me a few months before writing these pages. I wrote it down in my notes at the time. In retrospect I find it important that the theme of violence, racial confrontation, and the silencing of the human voice, was structurally present in my being-in-the-world long before I chanced upon the photograph on the Metropolitan Museum website. I naïvely follow Descartes who, according to his biographer, "attribuoit cette

"I am taking a shower in a place that looks like my fitness center. Next I am walking out but suddenly it comes to mind that I must have left a hairdryer in the showers where other people are also showering. I would like to go back and get it but I don't. It's no big deal. Then I am leaving a building where I just gave a lecture. It's late afternoon or evening. I am supposed to go home by bicycle. Somewhere else my friend and I find ourselves caught in a violent confrontation between white men, certainly colonial oppressors, and natives who look like indigenous people who are fighting back. One of the white men is hiding but he is soon discovered. He is going to be killed. My friend and I watch the scene covered by a large sheet of burlap. We are very uncomfortable, surely it's going to be gruesome, but we feel safe. We are not going to be involved or attacked. The white man is cornered and his mouth and face slowly covered with a thick, black liquid, like tar or pitch." The trauma of the dream persisted, thus pushing me to raise the question of the speaking voice being silenced. In hindsight I think the dream manifests a hermetic quality in silencing the blasphemous voice of the white man, and the black substance might hint at the *nigredo* phase of the alchemical process (the stage of blackness in the crucible). The killing did not happen in the dream contrary to our expectation, which was certainly vivid at one point. It remains disturbing to me that the white man's voice must be silenced. There is no revengeful killing, only the silencing of the impious colonialist. I wonder if the stifling of the oppressor's voice alludes to the incarnation of the racial terrorism, in words and deeds, that Christian Western civilization has visited upon the world. Humanism, rarely at peace with Christianity, undermined the metaphysical tradition that had established the primacy of truth (*veritas*) in the sense of *adequatio intellectus et rei*. For, humanism discovered that language is not the polished reflector of reality, but rather the response that being-in-the-world deploys to the calling of the situation. The relationship between words and things is highly variable in different situations. There is a contextual mind that constitutes being-in-the-world's mode of response to the call, and it is prior to the establishment of the possibility of adequation of mind and world. The humanistic fables chosen by Curione and Galileo, both magnifying the providential order operating in the insect microcosm, respond to the call of the hidden world, the call of the being-there, displaced, alongside an alien

---

merveille à la divinité de l' Enthousiasme, et à la force de l' Imagination, qui fait sortir le semences de la sagesse (qui se trouvent dans l'esprit de tous les homes comme les énticelles de feu dans le cailloux) avec beaucoup plus de facilité et beaucoup plus de brillant même, que ne peut faire la Raison dans les Philosophes" (Baillet 1691, p. 84).

mode of life, which is in the tradition of humanism's opposition to the criterion of truth as adequation. The insect does not care about adequation of mind and world. There is a reason, *ratio*, a mode of being in everydayness and of discovering one's at-homeness that properly belongs in both the spider and the cicada, and which transcends the dualistic schema of the metaphysics of adequation. The cicada that the inquirer kills is not recalcitrant to the scientist's epistemically motivated violence that strives, in vain, to determine the adequation of mind and world by killing. The insect is obedient. It is the final killing that reveals the true nature of the call from the abyss of the alien mode of being. This truth is not adequation of mind and world, the unfoldment of the mechanism that produces the harsh sound. Its vibration is a resonance from the 'beyond' of the multiplicity of beings, a revelation in a special sacred code, that reverberates in the voice of the cicada until its unique mode of existence is annihilated. The fable of the cicada hints at experience. But what is narrated by Galileo is not a sequence of trials and errors. It is not an experience in the modern sense of controlled experiment in the laboratory. It is an experience of revelation in nature. It is the obedience of nature that manifests the incarnation of the creative work of God. Being-in-the-world reveals itself authentically in an experience that responds to the call of the 'there' in nature. Both nature and Scripture are open for empirical pursuit by non-violent being-in-the-world. Both are parallel modalities of being-in-the-world of non-violent hereticality. The modality exemplified by the humanistic fable constitutes being alongside nature. The modality exemplified by Scriptural hermeneutics constitutes being alongside God.

The Scriptural modality of non-violent hereticality was defined by Fausto Sozzini. Here I follow Cantimori's acute exposition of the predicament that led Fausto to the formulation of a revolutionary theory of both Scriptural hermeneutics and Christian ethics in terms of 'sensata esperienza'. It was a striking precursor of the Galilean 'sensata esperienza', though the latter has been reduced by positivist historiography to Galileo's theory of knowledge as experiment. Cantimori's startling conclusion is that Sozzini's predicament was "un tentativo di risolvere i problemi al lume di una esperienza sensata (poiché per il Sozzini lo studio della Scrittura e la prassi della *imitatio Christi* avevano il valore di un'esperienza), che costitutiva una nuova *via* per intendere i doveri del cristiano".[31] It must be emphasized that it was both Scriptural hermeneutics and the praxis of *imitatio Christi* (imitation of Christ's life), involving

31 Cantimori 1992 (1st ed. 1939), p. 358. "...an attempt to solve the problems in the light of sensory experience (since, in Sozzini's view, the study of Scripture and the practice of

the humanistic domain of language with Christian ethics, that were constituted by Fausto's hereticality as *experience*. Scripture is as enigmatically open to experiential learning as Christ's life insofar as we can imitate the ideal of his being towards death. Being-in-the-world of non-violent hereticality finds itself alongside two exceptional entities that do not manifest themselves as anonymously graspable objects. One entity is nature, namely, divine revelation through the works of creation, and the other is the humanity of Christ (Sozzini rejected the divinity of Christ), the significance of whose precepts must be understood by humanistic exegesis of Scripture. Both manifest the character of entangling the humanness of being-in-the-world in experiences that authentically respond to the call of the beyond. Both share the character of an experience in the limit. The difference between Christ's nature and human nature remains subtly ambiguous, however. Cantimori claims that it is not so much qualitative but quantitative.[32] According to him, Fausto's anti-Trinitarianism aims at elevating humanness in a moral sense that opposes the Neoplatonic exaltation of the dignity of man, the emphasis on the immortality of soul, promulgated by Ficino and Pico. In fact, Fausto thought that the human being, body and soul, is by nature mortal, and that immortality and hence salvation is but a gift granted by God to humans for their faith and works guided by the imitation of Christ.[33] Fausto proposes a method for reaching eternal salvation that is experiential through and through. I recall that reenactment of Galileo's experiments demonstrated that, for instance, pendulums show isochronism (sameness of periodic time of oscillation regardless of amplitude of oscillation), synchronism (two pendulums keeping perfect pace), and discrepancy (the tendency to diverge from isochronism) along a continuum of patterns of behavior, none of which appears to correspond to modern expectations of conformity of theory and empirical outcome. There is no such thing, then, as the pendulum phenomenon as a graspable object outside human being-in-the-world. The pendulum has the character

---

imitatio Christi had the meaning of an experience), which was a new *method* for understanding Christian moral life".

32 Cantimori 1992 (1st ed. 1939), p. 355.

33 "Anzi, il Sozzini riteneva che l'uomo, —l'uomo intiero di anima e di corpo— fosse per natura di condizione mortale, e che l'immortalità fosse un dono di Dio da conquistarsi con la fede […] Cosí Fausto Sozzini rielabora i concetti della *Imitazione* di Cristo e della giustificazione della fede fondendoli in uno attraverso quello, tipicamente valdesiano, della giustificazione per la fede e per le opere; e attraverso l'esigenza della critica filologica dell' umanesimo applicata alle questioni teologiche fa della religione un metodo…". Cantimori 1992 (1st ed. 1939), pp. 356–357.

of an experience in the limit for being-in-the-world, a phenomenon which belongs in nature in its infinite processes, in the sense that isochronism and synchronism tend to manifest themselves better and better as the parameters that control the outcome of the experiment tend to certain values, but without ever reaching the ideal limits. More generally, this 'experience in the limit' is the hallmark of Galileo's famous experiments. They escape modern expectations of perfect matching between theoretical prediction and measurement because they are not experiments in the sense of the logical-positivistic conception of science (testability and meaning of a physical theory through empirical verification). In being alongside entities such as nature and the humanity of Christ, then, being-in-the-world of hereticality projects its ownmost possibility of understanding as salvation and hence immortality. Galileo's experiments are indeed experientially 'methodic' but cannot be dissociated as natural-philosophical events from his being-in-the-world as projecting his ownmost possibility of understanding nature towards immortality. Galileo's audacious pronouncement on the affinity of human intellect and God's intellect can be placed in this context of non-violent hereticality, yet it says something even more startling than Fausto's accentuation of the humanity of Christ.[34] The human intellect understands geometrical propositions as 'intensively' as God, with the same degree of certainty because it can understand their innermost necessity, above which there cannot be greater certainty for humans. But it cannot have access to all of them because they are infinite while our humanness is finite. As Christ is human and not divine, and thus imitable but only to a certain extent, because the experience of Christ is the limit of human experiential learning, so the intellect of God is imitable by the geometrician who grasps the essence of a geometrical truth and in this way

34 "Molto acutamente opponete; e per rispondere all'obbiezione, convien ricorrere a una distinzione filosofica, dicendo che l'intendere si può pigliare in due modi, cioè *intensive* o vero *extensive*: e che *extensive*, cioè quanto alla moltitudine degli intelligibili, che sono infiniti, l'intender umano è come nullo, quando bene egli intendesse mille proposizioni, perché mille rispetto all'infinità è come uno zero; ma pigliando l'intendere *intensive*, in quanto cotal termine importa intensivamente, cioè perfettamente, alcuna proposizione, dico che l'intelletto umano ne intende alcune cosí perfettamente, e ne ha cosí assoluta certezza, quanto se n'abbia l'istessa natura; e tali sono le scienze matematiche pure, cioè la geometria e l'aritmetica, delle quali l'intelletto divino ne sa bene infinite proposizioni di piú, perché le sa tutte, ma di quelle poche intese dall'intelletto umano credo che la cognizione agguagli la divina nella certezza obiettiva, poiché arriva a comprenderne la necessità, sopra la quale non par che possa esser sicurezza maggiore" (Galilei 1890–1909, VII, pp. 128–129).

it imitates God's cognition perfectly. In imitating the experience of Christ's life we tend towards the same intensity of understanding as Christ's because we experience, as humanly as Christ did, the necessity of our deeds as moral obligation towards eternity.

However, Galileo pushed the possibility of the human intellect beyond the model of imitation of Christ's life that Fausto had elaborated. For Fausto, when somebody wishes to imitate another it is enough that there appears to be a similarity between the qualities that both must express in relation to the thing that they are concerned with, so that both can be said to have qualities of the same genus even though the one who is being imitated by the other has the qualities in much higher degree and perfection.[35] In the parlance of scholastic philosophy intellect is a habit of mind, that is, an accident of substance falling under the category 'quality' (the first species of quality). In Galileo's model of imitation, we can see more than a similarity between shared qualities insofar as there is adumbrated the same degree of perfection in understanding geometrical truth in both the divine and the human intellect (though they follow radically different pathways to form this habit of the intellect, as Galileo hastens to point out).[36] Galileo integrates geometrical knowledge into the being alongside nature that frees the possibility of understanding the infinite operations of God, and hence correctly understanding the precepts that the imitation of Christ must obey. Geometry is a pathway towards illuminating knowledge, the gnosis that being alongside nature reports homewards. Nature is not present to us as a graspable object sitting outside of our being-in-the-world. Objective presence is the intellectual alibi by which we have given

35 "Quandoquidem, ut quis aliquem imitari dicatur, satis est illius praecipuas qualitates in ea re, de qua agitur, ita exprimere, ut similitudo inter utrumque appareat, et uterque eiusdem generis qualitates habere dici posit; licet is, quem alter sibi imitandum proposuit, multo perfectiores eas praestantioresque habeat". Quoted in Cantimori 1992 (1st ed. 1939), p. 355.

36 "Però, per meglio dichiararmi, dico che quanto alla verità di che ci danno cognizione le dimostrazioni matematiche, ella è l'istessa che conosce la sapienza divina; ma vi concederò bene che il modo col quale Iddio conosce le infinite proposizioni, delle quali noi conosciamo alcune poche, è sommamente piú eccellente del nostro, il quale procede con discorsi e con passaggi di conclusione in conclusione, dove il suo è di un semplice intuito: e dove noi, per esempio, per guadagnar la scienza d'alcune passioni del cerchio, che ne ha infinite, cominciando da una delle piú semplici e quella pigliando per sua definizione, passiamo con discorso ad un'altra, e da questa alla terza, e poi alla quarta etc., l'intelletto divino con la semplice apprensione della sua essenza comprende, senza temporaneo discorso, tutta la infinità di quelle passioni; le quali anco poi in effetto virtualmente si comprendono nelle definizioni di tutte le cose, e che poi finalmente, per esser infinite, forse sono una sola nell'essenza loro e nella mente divina" (Galilei 1890–1909, VII, p. 129).

ourselves permission to exploit and toxify the planet. Being alongside both nature and the humanity of Christ opens up possibilities of understanding by disclosing the subtle pathways that we can follow to effect the transformative power of imitation. Galileo insistently demarcated natural philosophy as the domain of the naturally knowable through geometrical reason and experience, but experience embraces more than watching the sky, or observing pendulums swinging and bronze balls rolling down inclined planes. The latter are the life-less remains of experience that positivism has severed from the existential infinity that we must restore. It is this severance that turned positivism into the hatred ideology that became the breeding ground for Western colonialism and racial terrorism. Galileo's descendants are not the Western scientists acclaimed by the regurgitated capitalist propaganda that nowadays passes for science education, but the African-American heretics expressing authentic sympathy in a hostile world.[37] The people massacred in Kentucky by white racial terrorists should take center stage in a future pedagogy of reparation that returns being-towards-death at the forefront of education. Fausto argues that God created humans naturally mortal as all creatures, but that after the Fall humans became subject to death because the grace of immortality had been revoked, otherwise it would be inconceivable that because of the lapse of the human species the entire non-human life on earth would have to become subject to the punishment of death. Death is neither a nature, nor a quality, nor the property of a thing, but rather the present or future abolition

37 The killing of the cicada is the archetype of the *species terrorism* that Western science ritually enacts in laboratories all around the world in which non-human animals are brutalized. Here is an example of the 'objective' language by which science ritualizes the brutality. "Squirrel monkeys weighing between 450 and 800g were anesthetized with Diabutal (sodium pentobarbital) at a dosage of 28 mg/kg. A tracheotomy was performed and a large portion of skull was cleaned of all tissue and cemented to a head holder. [...] The surgery was traumatic and many of the animals died. If the animal survived the surgery, the preparation usually remained quite stable for hours. In the later experiments very few animals died. Of a total of nearly 150 animals approximately 25 died; many of the others were used mainly to develop the experimental technique" (Rhode 1971, p. 1218). The rigorous absence of reflectivity, self-consciousness or commentary of any sort on the part of the scientist-priest writing the report about how the sacrifice of the monkeys has been correctly executed (note the obsessive passive voice at all junctures where violence is carried out erasing personal responsibility) is the dialectical transformation of the rituals that science appropriated from early modern Christianity, especially the Catholic mass, whose focus is the reenactment of the supreme sacrifice, Christ's crucifixion, the killing of the human-animal-God, during which the priest performing the miracle of transubstantiation recites objective formulaic language without any personal commentary.

of all of these ontological determinations. Death is not really the natural process of dissolution, corruption, but rather the unnatural death, premature, or due to miserable conditions, or occurring in any *violent* manner.[38] Hence it is being-towards death that naturally occurs to all creatures as the ultimate abolition or accomplishment of nature, quality, and thing, that Fausto believes is the essence of the creature. Galileo's *experience* follows the humble pathway of an imitation, a temporal pathway, to be sure, in sharp contrast to God's outstanding a-temporality (God's mode of knowing being, says Galileo, *un semplice intuito* of all truths, contemplated by God with equal clarity beyond time). Galileo's *experience* is a modality of being-in-the-world of non-violent hereticality. It decries the killing of the cicada, its violent death, as the barbarism of knowing for the sake of knowing. It is being-in-the-world alongside nature and the humanity of Christ, or, shall we say, alongside God and the divinity of the human being, intended *intensive*, not *extensive*.

Being-in-the-world of non-violent hereticality includes another fundamental modality. I am referring to biblical hermeneutics. In preparation for the *Eretici*, Cantimori edited (with Elisabeth Feist) an anthology of rare texts that he had collected in European libraries and archives, in which for the first time Sebastian Castellio's *De arte dubitandi* was published in a more complete form that had hitherto been known (but a few chapters were not reprinted as they had been published before).[39] Although Castellio was not of Italian origin, Cantimori demonstrated that he belonged in the same spiritual milieu as the Italian heretics. Castellio's *De arte dubitandi* is of great importance for his original approach to biblical hermeneutics. Galileo's involvement with the interpretation of Scriptures appears to be consonant in some respects with Castellio's hermeneutics. It is this harmony that I would like to pursue here as it has not been emphasized so far.

---

38 Sozzini 1610, pp. 10–12. I find Fausto's discovery of the naturalness of death, intended as final suspension of all ontological determinations, as opposed to violent or unnatural hence *real* death, as momentous as Galileo's astronomical observations with the *perspicillum* (telescope) that were published in *Sidereus nuncius* the same year as Fausto's *De statu primi hominis ante lapsum disputatio* (1610). There is in both writings the same spiritual effervescence inspired by a deeply-felt sense of the sacredness of human experience. Cf. Scribano & Priarolo 2006, for a collection of essays on Fausto Sozzini's influence on Socinianism and European philosophy. On Sozzini's philosophy, see Scribano 2006.

39 Castellio's *De arte dubitandi* is in Cantimori & Feist 1937, pp. 307–430. A second edition was published by Feist a few decades later with a newly written introduction (Castellio 1981).

I begin from the fact noted by Castellio that Scriptures are often obscure in regard to things that do not seem to pertain to salvation, such as baptism, the Holy Supper, justification, predestination, and others.[40] For Castellio, however, obscurity is not an accident. Indeed, it is not a negative feature of Scripture, and there is nothing absurd in the fact that God wanted the holy writings to be obscure about certain questions since it must be conceded that God wants nothing absurd. The reason why, according to Castellio, there is lingering in Scripture this persistent obscurity is that God wanted human beings to exercise their *ingenium* (inventiveness) in finding out the veiled meanings of Holy Writs instead of remaining passively idle.[41] The suggestion that God wants us to exercise *ingenium* is paramount. In the humanistic tradition which framed Italian hereticality, as Cantimori has shown, *ingenium* played a fundamental epistemic role. It means ingenuity, inventiveness, figurative language, poetry, fables, not the dogmatic application of rules of logic. The humanistic art and science of invention, the exercise of *ingenium*, explores the nature of human beings in their historical situatedness through the primacy of language as opposed to the primacy of being that is the presupposition of scholastic metaphysics. Castellio states that, since there is this obscurity that has been the cause of endless controversies, something must be found in which truth is so glaringly evident to everybody that it cannot be denied. This can only be *sensus* and *intellectus*, which are the instrumentalities of *ingenium*.[42] However, controversies that exceed the power of sensus and *intellectus* cannot be tackled by this method. Castellio claims that, since there are things which can be apprehended by *sensus* and *intellectus*, that is, that implicate the operation of the senses and reason in the processes of human understanding, whenever an interpretive controversy arises regarding things that fall under the purview of sensory apprehension, it must be resolved by *sensus* and *intellectus*. But what is *intellectus*? What is reason (*ratio*)? Reason, says Castellio, is that daughter of God that existed before God's creation of the universe and will exist forever after the final events that renovate the world. It is the eternal discourse of God (*sermo dei*), more ancient and more certain than all rituals and scriptures,

40 "Neque vero, obscuras dum esse dico sacras literas, sic obscuras dico, ut in eis nihil sit apertum. Sunt enim et quidem plurima aperta et in primis ea, quae ad hominis officium et porro salutem pertinent [...]. Sed obscura certe sunt ea, de quibus vulgo solent esse controversiae [...]. At de baptismo, de coena Domini, de iustificatione, de praedestinatione et caeteris multis dissensiones sunt capitales. Cur? Nimirum quia haec obscure nobis in divinis literis sunt tradita" (Cantimori & Feist 1937, p. 355).

41 Cantimori & Feist 1937, p. 356.

42 Cantimori & Feist 1937, p. 357.

by which God has instructed and will always instruct his people about divine things.[43] There are in Scripture many passages that are figurative. They must be understood correctly in the light of all literal passages that are laid open to immediate understanding. An example of figurative statement in the Bible is that Christ resurrected. It must be intended, according to Castellio, in the sense that he changed his life from bad to good, so that we understand that he was almost changed from being dead to being alive.[44] It is implied by Castellio, then, that figurative passages are such that they escape the possibility of immediate apprehension by human *sensus* and *intellectus*, so that their truth must be discovered by exercising *ingenium*, the activity that God wants us to engage in naturally, and finally by having recourse to eternal reason. But this eternal reason is itself embedded in discourse, language, it is the eternal discourse of God, by which he speaks to his people from his non-temporal perspective. There is certainly an interesting tension here, between the appeal to the sensory world that operates as the criterion for demarcating figurative from literal language, thus warranting the discrimination between what can and what cannot be grasped through the human senses, and the non-temporal perspective of God's eternal discourse that seems to transcend the historically situated human being grappling with the sensory world. Castellio eventually formulated the following hermeneutic rule. If any statement whatever is found in both holy and profane writers that is manifestly repugnant to the senses and to reason unless it is interpreted figuratively, then it must be interpreted figuratively in order that it can be reconciled with the senses or with reason.[45] Thus, our being-in-the-world displaced from the center of understanding, the self, our finding ourselves alongside nature, in a state of unresolved ambiguity as to the demarcation of the literal and the figurative, the immanent and the transcendent, was discovered by the humanist Castellio as the natural condition of existence, which is ontologically prior to the formation of the

43 "Nam ratio est ipsa, ut it loquar, dei filia, quae et ante literas et cerimonias omnes atque adeo ante orbem conditum fuit et post literas et cerimonias omnes atque adeo post mutatum novatumque hunc mundi statum semper futura est neque magis quam ipsemet deus aboleri potest. Ratio, inquam, est aeternus quidam sermo dei longe tum literis tum cerimoniis et antiquior et certior, secundum quam deus suos et ante cerimonias et literas docuit et post easdem ita docebit, ut sint vere divinitus docti" (Cantimori & Feist 1937, p. 363).

44 Cantimori & Feist 1937, p. 414.

45 "Et omnino generalem hanc regulam teneamus, si quod dictum vel in profanis vel in sacris authoribus eiusmodi est, ut, nisi figurate accipiatur, manifeste rationi aut sensibus repugnet, id esse figurate accipiendum. Itaque interpretandum, ut cum ratione aut sensibus concilietur"(Cantimori & Feist 1937, p. 419).

subjectivity that establishes the polarities of figurative and literal. The latter are a reflection in subjective understanding of the separation of self and world that is installed after the problematic hermeneutic situation has forced our being-in-the-world alongside nature to turn away from nature and hence differentiate a subject from an object of understanding.

The events leading to Galileo's infamous abjuration of 1633 and the motivations lying behind that episode have intrigued scholars for centuries. Such an erudite effort notwithstanding, there is still strong disagreement about virtually every detail concerning those events and their motivations. It is of course beyond the scope of this book to delve into the huge literature that has accumulated during the years since 1633. What interests us here is Galileo's approach to the relationship between scientific truth and biblical truth. One unresolved question regards Galileo's attitudes toward the inerrancy of the Bible and the potential conflict between Copernican astronomy and a traditional reading of some biblical passages, most notably the story from the *Book of Joshua* concerning Joshua's commanding the sun to stand still. In a number of manuscript writings in defense of Copernicanism, particularly in the well known *Letter to Grand Duchess Christine of Lorraine*, which he widely circulated at the time, Galileo re-proposed a hermeneutic principle, a first principle, for the interpretation of the Bible and put forward two carefully crafted examples of the application of that principle.[46] According to Galileo's formulation of the principle, *two truths can never contradict each other.*[47] Thus, for him, the truths attainable within natural philosophy by means of sensory experience and/or necessary demonstrations can and must be attuned to the infallible truths contained in Scripture, and vice versa. The hermeneutic first principle posited by Galileo was not formally deducible from higher principles. It was understandable to Galileo only within the context of his *ethical* experience of both Scripture and Copernican astronomy. The situation of ethical experience corresponds to Castellio's discovery of a hermeneutic mode of existence that is prior to the separation between subject and object of understanding. Ethical experience is a modality of being-in-the-world of

46 Those writings have all been published in the *National Edition* by Antonio Favaro (Galilei 1890–1909) under the heading *Scritture in difesa del sistema copernicano* [Writings in defense of the Copernican system]. The most elaborated is the 'Letter to Christine of Lorraine' on the relationship between science and religion. See the Italian originals in Galilei 1890–1909, V, pp. 261–370.

47 "Stante questo, ed essendo inoltre manifesto che due verità non possono mai contraddirsi tra loro…" (Galilei 1890–1909, V, p. 283).

hereticality that does not break the unity of our emotional apprehension of the world, our being alongside nature, and our knowing the world in self-reflectivity. This modality is the legacy of Italian humanism. The latter resisted the tendency, typical of scholastic philosophy, to equate knowing with deductive science. In this humanistic tradition, forms of knowledge are privileged that are not reducible to formal models of deduction from first principles. Since first principles are not demonstrable by definition, humanists tend to look for conditions that make knowledge of first principles accessible to mind by investigating natural languages. They especially emphasize the cognitive dimensions of poetry and rhetoric, that is, figurative and emotionally charged language. Humanistic rhetoric is not so much the art of persuasion as being-in-the-world that makes it possible for humans to articulate in words ethical experience. It is especially metaphor that has the power to reverberate ethical experience.[48] Hence the question of the demarcation of the literal and the figurative in language is prominent in humanism.

Galileo gave suggestions which were intended to show in practice how it was possible to reconcile the Bible with Copernican astronomy. The most technical example, first circulated in a letter to his pupil Bendetto Castelli (1613), concerned the disputed passage from the story of Joshua, whose command to the sun to stand still seemed to imply the motion of the sun. Although he claimed that in biblical interpretation the gravest of errors would be to insist on the literal meaning of words, for, taken literally, most biblical statements would be tantamount to blasphemy and heresy, Galileo ventured to explain the Joshua passage in a virtuoso fashion without attributing a figurative meaning to the story. He stated that what sensory experience puts before our eyes (a recurrent theme in Castellio's exegesis), or that which results from certain demonstrations, can by no means be doubted because of scriptural places that on the face of it appear to carry a discordant meaning. The reason for this priority of nature over Scripture is that while not every statement in Scripture is under the strictest obligation, nature cannot but observe her effects without exception. Then, Galileo repeats Castellio's argument that it is inconceivable that God, who has endowed us with senses, intellect, and reason, intended to give us by another means the information that we can gather with them.[49] In a final crescendo, he launches into his hermeneutic acrobatics by arguing that

---

48 Grassi 1988.

49 "Ma che quel medesimo Dio che ci ha dotati di sensi, di discorso e d'intelletto, abbia voluto, posponendo l'uso di questi, darei con altro mezzo le notizie che per quelli possiamo conseguire, non penso che sia necessario il crederlo…" (Galilei 1890–1909, V, p. 284).

in Ptolemaic astronomy the sun's proper motion, the only one that pertains to it, runs in the opposite direction to the motion of the *primum mobile*, the prime mover, that sets the twenty-four hour, daily rhythm of the universe. So Joshua's command would have been totally ineffective, indeed counterproductive since its outcome would have shortened the day slightly instead of the expected lengthening. Contrariwise, according to him, in the Copernican system the command can be interpreted as invoking the cessation of the polar rotation of the sun on its axis, a motion that Galileo proudly attributed to the solar body on the basis of his discovery of the rotation of sunspots.[50] It is a hermeneutic feat of literalism that must have raised the eyebrows of Cardinal Bellarmino and the Catholic theologians. Ethical experience as being-in-the-world of hereticality preserves the unity of our being alongside nature and our knowing the world in self-reflectivity. The first aspect of this ontological unity prior to the distinction of subjectivity and objectivity is signaled by Galileo's literalism in the case of Copernican astronomy. His solution to the dilemma of the indeterminacy of the borderline between literal and figurative is an exorbitant extension of the domain of the literal to encompass both astronomy and the biblical story. The second aspect is the disclaimer that unlike the Bible nature is observant of her law-like necessity in the strictest possible sense. No effect that *sensata esperienza* discloses before our eyes can be doubted, while the exposition of the Bible remains open to further investigation in the light of the inexorability of natural-philosophical truth when it has been certified by the senses or demonstrated by geometrical reason. Being-in-the-world of hereticality alongside nature projects radical possibilities of understanding that once again direct us towards the problem of philosophical violence. This modality of being displaces the self from the orthodox center, as it pertains to an understanding of nature through the language of geometry

50 "Avendo dunque io scoperto e necessariamente dimostrato, il globo del Sole rivolgersi in sè stesso, facendo un'intera conversione in un mese lunare in circa, per quel verso appunto che si fanno tutte le altre conversioni celesti; ed essendo, di più, molto probabile e ragionevole che il Sole, come strumento e ministro massimo della natura, quasi cuor del mondo, dia non solamente, com'egli chiaramente dà, luce, ma il moto ancora a tutti i pianeti che intorno se gli raggirano; se, conforme alla posizion del Copernico, noi attribuirem alla Terra principalmente la conversione diurna; chi non vede che per fermare tutto il sistema, onde, senza alterare il restante delle scambievoli relazioni de' pianeti, solo si prolungasse lo spazio e 'l tempo della diurna illuminazione, bastò che fusse fermato 'l Sole, com' appunto suonan le parole del sacro testo? Ecco, dunque, il modo secondo il quale, senza introdur confusione alcuna tra le parti del mondo, e senza alterazion delle parole della Scrittura, si può, col fermar il Sole, allungar il giorno in Terra" (Galilei 1890–1909, V, p. 287–288).

and emotionally charged experience that prevent the self from relinquishing the assent that being alongside an entity not graspable as an object forces as absolutely ineluctable. Both nature and Scripture manifest this non object-like character. Galileo emphasizes the sense of ineluctability that being alongside nature provokes. He underlines that the demonstrative sciences constrain assent in a manner that cannot be negotiated. To deny assent to the demonstrative sciences would require learning how to manipulate the faculties of the soul, so that the will and the imagination might overpower the intellect.[51] It is a fascinating insight into the secret play of philosophical violence against the existential freedom of hereticality. The powers of the soul cannot be resisted, insists Galileo. To further the position that biblical hermeneutic must take priority over the demonstrative sciences is tantamount to raping the soul. Being-in-the-world of hereticality alongside nature finds itself displaced from orthodox anonymity and recognizes that its ownmost authenticity speaking through soul is ineluctable. Italian hereticality discovered this possibility of authentic existence in being alongside entities that cannot be grasped as manipulable objects by a subject standing alone in opposition to a world severed from its interpretative statements. The discovery was powerful enough to derail Galileo into a mortal confrontation with the Roman Catholic Church, though he finally retracted. As Mauro Pesce has pointed out, Galileo's heretical hermeneutics was not acceptable to the Catholic hierarchy that eventually summoned him to stand trial in Rome.[52] But neither was his claim to the ineluctability of the heretical soul so inflexible as to make him stand firm in the face of violent death.

51 "Il comandar poi a gli stessi professori d'astronomia, che procurino per lor medesimi di cautelarsi contro alle proprie osservazioni e dimostrazioni, come quelle che non possino esser altro che fallacie e sofismi, è un comandargli cosa più che impossibile a farsi; perché non solamente se gli comanda che non vegghino quel che e' veggono e che non intendino quel che gl'intendono, ma che, cercando, trovino il contrario di quello che gli vien per le mani. Però prima che far questo, bisognerebbe che fusse lor mostrato il modo di far che le potenze dell'anima si comandassero l'una all'altra, e le inferiori alle superiori, sì che l'immaginativa e la volontà potessero e volessero credere il contrario di quel che l'intelletto intende (parlo sempre delle proposizioni pure naturali e che non sono de Fide, e non delle sopranaturali e de Fide)" (Galilei 2000, p. 113.)

52 *Introduction*, in Galilei 2000, pp. 9–66.

Libertinism

Italian libertinism of the (late) sixteenth and seventeenth centuries is a paradoxical historiographic category. The most important work on it remains Giorgio Spini's *Ricerca dei libertini.*[53] (I emphasize that I will limit my discussion to the phenomenon of Italian libertinism, as the wider ramifications of libertinism in Europe are beyond the scope of my research. Thus, I will use the word 'libertinism' in a restricted sense, referring to the Italian phenomenon that was studied by Spini in his ground-breaking monograph). Spini defined libertinism as nothing but a generic, negative category. It is not so much a philosophical school, or the beginning of a renovation, as the bursting into a peal of laughter at credulity, it is rebelliousness, intolerance of dogma and morality, a penchant for scandal and impertinence. It is not a set of doctrines, but a state of the soul (*stato d'animo*), indeterminate, which is poised on the verge of nothingness, prone to degenerate into vulgarity, libel and pornography.[54] There is no doubt that Spini had a very pessimistic view of libertinism, a spiritual illness infecting the early modern age, a nightmare, the unconscious of an epoch of the spirit that must be brought to light in order to understand the past and the present. However, what motivates my interest in Spini's unique approach to libertinism is that in his doing the work of historical analysis, or rather psychoanalysis, unearthing the hidden sinfulness of our historical consciousness, as he put it, his pessimism turns into a powerful existential mode of inquiry that involves the writer's own consciousness into the process of hermeneutics.[55] This is made explicit in his preface to the second edition of the book, where Spini delineates the state of soul in which the project was originally conceived, that is, under the influence of the political events following the fall of Fascism in Italy, the dramatic conclusion of second world-war, and the irreconcilable splitting of the emerging political

---

53 Spini 1983 (1st ed. 1950). I will follow in my exposition the second edition of the book (1983), which has an important "Postilla Galileiana" (*ibid.*, pp. 389–392), an entirely reworked chapter on Giulio Cesare Vanini, and an illuminating autobiographical preface.

54 "Giacché libertinismo, di per sé stesso, non è altro che un' espressione generica, indeterminata, connotata soltanto per negazione. Non è tanto scuola filosofica o inizio di rinnovamento positivo, quanto sogghigno beffardo dell' altrui credulità, ribellismo generico, fastidio di dogmi e regole morali, gusto dello scandalo e dell' impertinenza. Non è affatto dottrina quanto stato d'animo, che appunto, come ogni stato d'animo indeterminato, rischia ad ogni momento di dissiparsi nel nulla, e di concludersi soltanto nella risata sguaiata, nel libello diffamtorio, nella pornografia". Cf. Spini 1983 (1st ed. 1950), p. 13.

55 Spini 1983 (1st ed. 1950), p. 13.

arena in Italy between on the one hand the ideology of catholic conservatism, returning to the spirit of the Counter-Reformation, and on the other hand the Marxism-Leninism-Stalinism of the Italian Communist Party.[56] The traumatic historical juncture transformed an erudite research into an original existential historiography that constantly challenged, throughout the book, the restrictive, pessimistic characterization that I mentioned above. Not that the pessimism was unjustified. On the contrary, Spini wrote magisterial pages on what he termed 'a century without apocalypses', devoid of eschatological tension. The *Seicento* in Italy had two souls, the Counter-Reformation and its secular double, impersonated by Renaissance naturalists such as the Averroistic philosophers at Padua, or by the figure of the philosopher-magus, Giordano Bruno, or Tommaso Campanella, to whom the Protestant principle of *sola scriptura*, in which Spini saw the motor of modernity, remained incomprehensible because it negated the mission of the high-priest-magus who was to legislate the advent of renovated Christianity over and against the ignorant populace. Spini denounced the slavery of early modern materialism and astrological determinism, the legacy of antiquity, and the inhumanity and humiliation to which the disadvantaged classes were subjected. But his most inspired and ruthless analyses concern the gender relationship. It is worth reading his conclusions. "The seventeenth-century custom, the mentality of the time in Italy, restored the masculine tyranny of paganism in its abysmal brutality and ferociousness. The Italian woman of the seventeenth-century is without exception a passive instrument of male caprice and calculation. Despicable courtesan, or forced nun, she must always retreat before the male, for he is the one destined to preserve the glory and wealth of the family, or otherwise she is the degraded object of mercantile negotiations among family groups, a veritable soulless thing".[57] Finally, Spini explained his assessment according to which Galileo was never a libertine in the sense that he had explored in the book. For, in his view, Galileo's life epitomized the scientist's heroic struggle

56 Spini 1983 (1st ed. 1950), pp. ix–x.

57 "Il costume seicentesco, viceversa, la mentalità corrente in Italia durante quel secolo, restaurano la tirannide maschile del paganesimo, in tutta la sua brutalità e la sua ferocia. Cortigiana spregiata o monaca a forza, per far largo al maschio, destinato a continuare in tutto il suo splendore la ricchezza della famiglia, sposa mercanteggiata dal suo gruppo familiare e trattata da marito, genitori, fratelli, come una cosa senz' anima, la donna Italiana del Seicento, è senza eccezione sensibile, lo strumento passivo del calcolo o del capriccio maschile". Cf. Spini 1983 (1st ed. 1950), pp. 48–50, quotation on p. 49.

for human progress.[58] The emergence of modern science, in other words, had nothing to do with heterodoxy, unbelief, or libertinism.

In what follows I appropriate and reformulate Spini's intuition that libertinism is not so much a doctrinal body as rather a state of the soul poised on the abyss of nothingness. Thus, I interpret libertinism as a modality of being-in-the-world that cannot be defined exclusively in doctrinal terms by specifying a corpus of heterodox theories. In fact, it is almost impossible to pin down any philosophical doctrine that might define the phenomenon of Italian libertinism. This is not to say that the Italian libertines do not hold certain philosophical or religious or political views, or that they are not seriously committed to their beliefs. Indeed, in some cases, such as Bruno or Vanini, they do not recant even in the face of death. The existentialist approach to libertinism entangles the interpreter in the process of interpretation in a profound sense. Spini's candid confession regarding the political climate that inspired the book bears testimony to this involvement. Libertines tend to distance themselves from the prevailing opinion, from the received wisdom of the age that stifles the mind. They share a deeper sense that it is the anonymous chorus of the others that wants to speak through the voice of the individual, and that the search for individuation must come to terms with the anonymity of the 'they' who have invaded and taken possession of the individual self. This coming to terms with one's individuation, this casting away from the anonymous 'they' is what being-in-the-world as if poised on the abyss of nothingness means. It is a form of understanding that is proper to being-in-the-world as it embraces the impulse to move away from the imposture of the anonymous 'they' and towards the emptiness that says "I". It cannot be accidental that there are common themes running through the story of libertinism such as the theory of religion as an imposture, or the attack on the Christian taboo surrounding sexuality. Where the wisdom of ages says 'Do not go', where the heart of darkness is kept hidden from view, there the libertine is irresistibly attracted to move. However, the libertine's struggle to move away from the 'they' does not attain the desired center, the focus from which a mysterious 'I' appears to radiate. The forbidden spaces that the libertine is

58 Spini 1983 (1st ed. 1950), p. 392. In the 1983 "Postilla Galileiana", Spini answered some critics that had solicited his views on Galileo. He pointed out that although it is true that, for instance, Galileo was for many years on very familiar terms with the most notorious of the Padua libertines, the Averroistic philosopher Cesare Cremonini, who was the godfather of unbelief in the Republic of Venice, Galileo did not share Cremonini's blind faith in Aristotle. Cf. also Spini 1996, pp. 23 ff.

attracted to are always defined by the anonymous 'they'. If these spaces are cordoned off by social custom or by theological and/or philosophical dogma, it is in virtue of a prohibition that is still enforced by the anonymous 'they'.

I would like to follow this thread of libertinism in Galileo's life-world by starting from his libertine manifesto, the *Capitolo contro il portar la toga*, a satirical poem that celebrates nudism.[59] David Wootton has emphasized that whatever we make of this poem, Galileo's poetic message is an overt attack on Christianity. In exalting nudity as an egalitarian form of life, in which all human beings are equal, and in condemning clothing (especially the academic gown) as an astute invention of the devil that has brought social inequality, Galileo seems to be advocating a return to the nakedness of Adam and Eve in the state of original justice, before the Fall, for then there will be no sin and hence no salvation.[60] In addition to the social-political utopia sketched by Galileo, the poem portrays sexuality as freedom from the nuisance of clothing, sarcastically denouncing the sexual impediments due to the academic gown. The obligation for the university professor to wear the *toga* in social settings prevents him from having fun with prostitutes at leisure, and instead forces him to have recourse to masturbation, which regrettably causes great consternation in his father confessor, who in turn worries about the professor's eternal salvation. Nudism, on the other hand, facilitates the selection of an appropriate partner and avoids the litigiosity ensuing inopportune marriages between disproportionately endowed people. The public scrutiny of the genitals, laid out for everybody to make an assessment, as we do in the market-place with goods that are up for sale in precise quantities and measures, encourages the choice of a proportional partner, so that 'what you see is what you get', no surprise, satisfaction guaranteed. Empowering the sexes to scrutinize each other's goods as buyers and sellers belonging in a market economy does not mean overtly renouncing the Christian ethics of marriage and the subordination of sexual pleasure to procreation. But it raises the question of the mechanics of procreation. It is the well-oiling of the machinery of sexuality that takes center stage in the marital relationship, and which may ultimately relegate marriage to a customary slot, a social convenience the usefulness of which

59 A *chapter against wearing the gown* (Galilei 1890–1909, IX, pp. 213–223). The genre of the poem belonged in the established tradition of satirical poetry that was popular in Italy in the sixteenth century (Reynolds 1982).

60 Wootton 2010, pp. 46–48. Spini plays down the poem. Though he acknowledges that it is reminiscent of the libertine attack on religion, he insists that it should be read as a jocular (*burlesco*) poem. See Spini 1983 (1st ed. 1950), p. 390.

becomes questionable in the utopia of nudity. Galileo adumbrates the nakedness of the genitals as the criterion of truth among equal human beings. Thus the move away from the anonymous 'they' of Christian sexual ethics takes Galileo-the-libertine towards the uncharted territory of nakedness. He places the genitals at the center of the cosmos, forging a Copernican ethics which tears the flesh of Christianity no less than heliocentric astronomy. There was an uncanny presence on the shelves of Galileo's library, the *De doctrina promiscua* by the humanist and doctor Galeotto Marzio (dedicated to Lorenzo de' Medici).[61] Marzio had written another work on anatomical terminology describing the parts (*membra*) of the human male organism from head to foot, in which he had stated that the virile penis (*membrum*) was the only *member* to deserve to be given that name, for its excellence as the organ of fecundity and procreation recognized since antiquity.[62] But in the *De doctrina promiscua* there is an apology of sexuality, a declaration of sexual liberation, that challenges Christian morality and the primacy of the virtue of chastity. Marzio presents an apologetic discourse of the transmutability of the male and female sexual organs that challenges the gender definition underlying the ethics of procreation, and explicitly aims at establishing that during sexual intercourse the female has greater pleasure than the male.[63] The argument is based on anatomical observation and supported by classical quotations. Marzio says that there is a certain similarity between male and female genitals. But the important difference is that while the penis is extended outside, the female penis is oriented towards the inside of the vagina so that the female testicles do not appear to be hanging outside. The male genitals can be changed into the female ones and vice versa by pushing them inside or outside. Sexual pleasure depends on the sense of touch. Thus, to determine whether the female or the male has the greater pleasure in sexual intercourse one has to consider the mechanics of touch. The male semen is hotter than the female semen, and generally the male has a more temperate disposition, from which a better friction ensues, and hence, it might seem, a more perfect titillation, that is, a higher pleasure for the male. But other factors must be considered. The

61 Miggiano 2008. See Favaro 1887, for a description of Galileo's personal library (p. 26, for Marzio's book).

62 "…membrum est, ob excellentiam sic appelatum. Nam cum plura in homine membra sint, hoc solum membri nomen optinuit, ut cum absolute dicatur, sit virile membrum. Hoc tribuit ei antiquitas propter foecunditatem, solum enim hoc generat" (Marzio 1517, p. 29).

63 Cf. Chapter XVIII, *De mulieribus in viros conversis, et maris an foeminae in coitu sit maior voluptas*, in Marzio 1552, pp. 185 ff.

highest pleasure is caused in the female because by receiving the hot male semen the womb is warmed up, but above all because of the friction of the penis inside the vagina. The penis titillates the interior of the vagina where the source of female pleasure is most concentrated. Furthermore, the body of the male is somewhat damaged during sexual intercourse as it discharges the semen, while that of the female is augmented by the reception of the male hot fluid. The correct proportionality of the size of the penis with respect to the opening of the vagina guarantees maximum pleasure in the female.

The upshot of Marzio's discourse is that sexual pleasure comes in a variety of wonderful flavors, and that changing sexual identity from male to female and vice versa, far from being absurd, far from being sinful, is in fact an anatomical predisposition, a wonder of creation, by which human beings can be considered gender neutral, androgynous creatures. Galileo-the-libertine's poem describes the utopia of nudism, hinting at the sexual liberation proclaimed by the humanist doctor Galeotto Marzio. It was an apologia for sexual Copernicanism against the Ptolemaic ethics of chastity upheld by Christianity. The genitals became the sun of pleasure at the center of the human cosmos. Galileo failed to put forward a convincing physical argument for the motion of the earth around the sun. The astronomical evidence of the cratered moon, the satellites of Jupiter, the phases of Venus, and the rotation of sunspots, did not deliver the decisive blow that could knock down the geocentric system of the universe. Even the phases of Venus, certifying that Venus rotates around the sun, could be accommodated by the Tychonic arrangement of the universe. The argument from tides, which, in his view, should have been the crowning achievement of the discussion led by Salviati in the 1632 *Dialogo*, remained problematic until the very end of his life.[64] I think that for Galileo Copernicanism was first and foremost a matter of pleasure, a secret incantation, the source of which was unconscious, for it was the aesthetics of the heliocentric system that must have pleased him, long before the news from Holland of a mysterious optical tube reached him in Padua. It was the utopia of an astronomical libertinism that put the sun, the cosmic *membrum*, at the center of the cosmological arena. He wrote Kepler in 1597 that he had been a Copernican for many years but he refrained from telling him his motivations, except that he alluded, cryptically, to the fact that he had succeeded in explaining certain phenomena that he thought inexplicable on the geocentric hypothesis. Further, he confessed to Kepler that he had no intention of making a fool of himself by publishing his findings since too many

64 Palmieri 1998.

idiots thought that the heliocentric system was bogus.[65] In 1597 Galileo did not know why he had long been a Copernican. He had no physical argument for the motion of the earth (unless what he had alluded to in the letter was an embryonic form of the argument from tides). The young Galileo's Copernicanism was not yet a consistent cosmology, but rather a libertine mode of understanding, a displacement, or perhaps the beginning of a displacement, an aesthetic move away from the 'idiocy' of the anonymous 'they' (...*tantus enim est stultorum numerus*) towards a 'no-go' area of the scientific-philosophical culture of the time.[66] The problem with the origin of Galileo's Copernicanism is that we might be tempted to see it as a conversion from belief in something to something different. It is a way of thinking that presupposes an understanding of the human being as a repository in which belief-objects are stored, and from which they can be removed at will. I do not see it this way. Being-in-the-world is prior to having beliefs. When Galileo's boasts to Kepler that he has written on the arguments pro and con Copernicanism we should ask ourselves what he really meant. This is not to accuse him of not telling the truth. It is simply to ask the question of what 'writing that is not intended to be public' means. Encountering objects such as the Copernican hypothesis with which one is not familiar brings the world of everydayness into relief. Being-in-the-world means that the world is not given as a separate object from one's subjective being and vice versa. Copernicanism is experienced

65 Galilei 1890–1909, X, p. 68.

66 Spini has raised fascinating questions concerning the mysterious origin of Galileo's Copernicanism. It is important to realize that we do not yet understand why and when Galileo became a Copernican. The philosophical prejudice that portrays Copernicanism as the 'right' theory blots out our possibilities of understanding its origin. Spini suggested that the Florentine current of Neoplatonism that had pervaded the culture of the Medici milieu since Marsilio Ficino was embodied in sophisticated policies of two Grand Dukes, Francesco I and his brother Ferdinando, which implied the sovereignty of the sun. For example, fortified cities such as *Eliopoli* (Città del Sole, a Tuscan enclave in the region of Romagna) at the frontiers of the Duchy were built or restructured, to which a name was assigned or other emblems symbolizing the universal power of the sun. Galileo's earlier interests in fortification theory is well attested by his manuscript treatises, and Florentine architects were involved in the construction of the city-fortress Palmanova, precisely when Galileo was permitted by the Medic to leave Pisa to take up his new teaching job at Padua. There is an interesting connection between Galileo's involvement with fortifications and Copernicanism which has not been scrutinized. Spini argued that the Medici's political heliocentrism was the seed-bed in which Galileo's cosmological opinions germinated. However, this political safety zone might have been too delicate for Galileo to come out as early as 1597. See Spini 1996, pp. 12–14.

by the young Galileo through a modality of libertinism. It need not imply a conversion simply because being-in-the-world need not have belief-objects in store. A persistent residue of this early libertinism might well be Galileo's apparently dogmatic rejection of Kepler's ellipse. Erwin Panofsky argued that it was Galileo's infatuation with the aesthetics of the sphere and the circle, a legacy of the artistic canons of the early Italian Renaissance, that prevented him from accepting elliptical astronomy into his own system of the universe.[67] That infatuation must have been grounded in the secret or unconscious pleasure that had gratified the imaginative withdrawing from the geocentric 'they' enacted by Galileo in his private writings. The anonymous 'they' is not an 'other' outside of being-in-the-world. It is structurally part of it, though it is not historically fixed. Withdrawal from the 'they' cannot be a severance from the community of others. It can only be a process of integration that is adaptive. The appropriation of the aesthetic canon of the circle is reinforced by the adaptive integration of the heliocentric hypothesis into Galileo's geocentric life-world. Another way of putting it would be as follows. Galileo's being-in-the-world was both geocentric and heliocentric at the same time, which sounds paradoxical for the champion of the scientific revolution. But there is no contradiction here because there are no belief-objects stored in being-in-the-world. The latter's nature is an openness towards the possibilities of authentic existence. The themes of aesthetics and pleasure lead us to biology.

The poem celebrating Galileo's sexual utopia should be seen in historical and existential continuity with libertine biology and the physiology of sex. For instance, Spini has commented on the radically anti-Christian attitudes towards sexuality of Giulio Cesare Vanini, against which it is instructive to project Galileo's being-in-the-world of libertinism. It is not that there is a genealogy of texts, or that the two figures might have been related, though perhaps Vanini studied at Padua when Galileo was there. But their life-worlds are not orthogonal either. Vanini's dialogs between Giulio Cesare and his friend Alexander on the hidden secrets of nature, published in Paris in 1616, are as vitriolic as the more famous exchanges between Simplicio and Salaviati in Galileo's *Dialogo*. Both authors wear masks to escape ecclesiastical or political persecution. Vanini's biology begins by demonstrating the spontaneous generation of mankind from the earth. But if the origin of the first human was once the result of the creative spontaneity of matter, why does the process not continue to happen all the time? Perhaps because of a unique astrological conjuncture that was once favorable to the generation of human life. Human

67 Panofsky 1954, pp. 20 ff.

forms could then be promptly united to matter so that the first human beings were born. But if putrefaction is the origin of life, why, insists the incredulous Alexander, are animals not being generated from putrefaction all the time? Others have dreamed, retorts Giulio Cesare, that the first humans were born of the putrefied matter of frogs, pigs, and monkeys, for humans are similar to these animals both in their flesh and in their customs, though some think the Ethiopians were born of the semen and genus of monkeys because of the color of their skin is the same. This is weird!, Alexander exclaims, how about the uniqueness of human upright posture? It is said by the ancient pagans that in fact the first humans were quadrupeds and that it is a matter of upbringing during which the limbs of children were opportunely constrained. An experiment should be done to get to the bottom of this question, suggests Alexander. A child might be raised in the jungle to see if it grows up naturally like a quadruped. Why was mankind created? There is prime matter, according to the Averroists, and there is God, pure actuality, and finally there are the immaterial substances in close proximity with God. Humans were created to be intermediate creatures between God and matter. Was mankind not supposed to have power over the animal kingdom, asks Alexander? Not at all! Humans have no control on animals, and as we prey on them, so it is more often the case that we are preyed upon by animals. Certainly, Alexander replies, but this happens because of original sin. Indeed, that was a most happy affair after all, Vanini concludes sarcastically, for now we have dominion over both animals and angels, and we can even summon the angels to be our pedagogues (as the theologians at the Sorbonne can do).[68] On the sense of touch and titillation, Vanini tells us that sexual pleasure is not due to the mechanics of touch, for if you touch the semen with your hands, where human touch is most perfect, you will nevertheless not feel any pleasure. Rather, sexual pleasure is a sixth sense, coming not from touch though touch is necessary. Sexual pleasure comes from spirit. Those who have prolonged sexual activities still feel pleasure even after they no longer ejaculate semen. Occasionally semen flows out without any pleasure, which means that it is not the semen that carries pleasure. Vanini's views were also based on anatomical observation. In Paris he saw a dissection in which a high concentration of semen was seen in the spermatic vessels, which, according to him, indicated that there is a concentration of semen under pressure in the testicles building up before ejaculation. He argues against Hippocrates that this anatomical observation demonstrates that the semen cannot come from all parts of the human body.

---

68 Vanini 1616, pp. 232–235.

The reaction of the whole body to sexual intercourse is not due to the affluence of semen from all parts of the body, but to the heat that is generated and which flows outside the body, so that the entire organism suffers a sort of enervation and relaxation. Next Alexander comes to the thorny issue of pleasure and procreation. Is the exquisite pleasure associated with sexual intercourse in view of procreation? Not at all, on the contrary, pleasure is maximized when there is no thought of procreation mixed up with sexual activity, for the intention of the agent must be the end of the action, and if the focus is on procreation, pleasure is diminished for during intercourse the procreation intended by the agent is not yet actually present. The quality of the sexual intercourse determines the quality of procreation. Thus, Giulio Cesare confesses that he wished he had been born of an extramarital relationship for then the heated passion of the lovers would have generated a wonderful child.

Unfortunately, his father was in his seventies, and his mother a disproportionately younger girl. The father had no longer the strengths, the high spirits, and his natural faculties were so debilitated that his semen was imbued with little force and hence feebly ejaculated. Otherwise, he would have been endowed with a superior elegance of form, robustness, force, and above all with a clear mind. Fortunately, Giulio Cesare admits, the adolescent mother was able to kindle the fading force of her decrepit husband.[69] On the other hand, philosophers are generally inept at procreation since their spirits are diverted from the heart towards the brain, so that the genitals are undernourished and therefore inefficacious. To the populace they may appear to be exceedingly libidinous only because they refuse to be subjugated to the demands of marriage, in which case their amorous relationships, though perhaps rare, are perceived as damnable. The fact is that the eyes of the masses are turned to them because of their illustrious name and fame, not to praise them, far from it, but in envious disparagement, and they have spectators in the manner of the sun, that is, only when they fall.[70] The philosophers-suns, in other words, do not mistake sexual pleasure for procreation although they know that they may be thwarted by their mental powers. Sexuality is the locus where the libertine is attracted to in order to seek authenticity. It is the eruption of pleasure in

69 Vanini 1616, pp. 311–324.

70 "Oculis quidem vulgi videntur sapientes esse libidinosi, quod matrimonii capistris ora praebere recusantes, ipsorum concubitus et si rari, damnabiles tamen existimantur, et conspicui fiunt, nam ob nominis splendorem, et magnitudinem, omnium oculi in ipsos convertuntur, non ad laudem, sed ad obtrectationem, ut merito Soli a nobis assimilentur, cui non nisi cum deficit, spectatores habet" (Vanini 1616, p. 320).

the sexed body that flings open the gates of libertine existence. Vanini's dialogic movement alternates between laughter and derision, praise and blame, always tasting the erotic pleasure of the incursion into the prohibited zone, crossing the threshold that demarcates heresy from Christian orthodoxy. Galileo's sexual utopia and Vanini's physiology of existence reflect the same erotic modality of being-in-the-world of libertinism. They share an emphasis on the rhetoric of praise and blame, or epideictic rhetoric, as the discourse that brings into relief the bodily-thoughts fueling the dialectics or pretend-play of altercation. Brian Vickers acutely noted that "the dominant rhetorical technique in the *Dialogue* [i.e., Galileo's *Dialogo*] is the simultaneous use of praise and blame, elevating the Copernican world-system and debasing the Ptolemaic".[71] Vickers insisted on the brilliant figure of the 'postiller', an anonymous voice that wryly comments on the discussion that is taking place from the margins of the pages, where Galileo added lapidary descriptions by means of which "[a] lthough limited by the conventional brevity of marginal notes, Galileo manages to give him [the postiller] a great range of commentary, from the mocking to the abusive".[72] Galileo-the-postiller takes the same erotic pleasure in apportioning praise and blame that he has tasted in the poem against wearing clothes. Galileo-the-libertine constantly jumps in and out of the written page, changing persona all the time, and passing from authorial self-consciousness (Galileo) to pretended anonymity (the postiller). He is taking the lead in occupying the public space of the 'they' by lending his often violent voice to the figure of the anonymous postiller. Epideictic rhetoric becomes libertine logic at the hands of Galileo and Vanini, an instrumentalist type of logic by which the libertine performs the comedy of philosophy on the stage of the world. It was at Padua that, in stark opposition to the late medieval theories of logic as a science, logicians and natural philosophers such as Giacomo Zabarella and his successor Cesare Cremonini had elaborated an instrumentalist conception of logic (meaning that logic was not itself a *scientia* but an instrumentality of reasoning in the service of the sciences).[73]

When Galileo published the *Dialogo*, Antonio Rocco was sucked in by the libertine game of praise and blame that was staged in the book. Rocco belonged in the *Accademia degli Incogniti*, a Venetian circle of libertine propaganda.

71 Vickers 1983, p. 71.

72 Vickers 1983, p. 96.

73 Cf. Zabarella 1608, columns 5 ff. His *Opera logica* was first published in 1578. It became the most influential corpus of early modern logical teachings, especially in Italian and German universities.

Its ideology was inspired by the heterodox teachings of the Averroistic school of Aristotelian natural philosophy at the University of Padua and by its most notorious master, the unbeliever Cremonini.[74] Rocco published a response to Galileo's *Dialogo*, the *Esercitationi filosofiche*, in which he rebutted Galileo's attacks on Aristotelian philosophy. Despised by Antonio Favaro, who published the *Esercitationi* in the *National Edition* only because Galileo-the-postiller immediately set out to write copious notes on many pages of the infamous book, Rocco's response engages the *Dialogo* on a par with Galileo, without sparing Galileo the same violent sarcasm that Galileo had poured on Simplicio.[75] Reading the exchange between Rocco and Galileo is like watching libertine logic coming alive. Neither Zabarella nor Cremonini nor Rocco did blindly subscribe to Aristotle's natural philosophy, a prejudice that we should do well to get rid of once and for all. They emphasized that philosophy must be based on first principles from which syllogistic deductions must be constructed whenever possible. This methodological precept is present throughout Rocco's *Esercitationi*. Galileo, a colleague of Cremonini for eighteen years, recognized this philosophical lineage as he underlined a passage in the preface of Rocco's book in which his opponent candidly confesses that he has scrutinized Galileo's novel principles, he has tried hard to apply them, but he has ultimately found them to be visions, illusion, and fantasy.[76] First principles are logical instrumentalities that must be put to the test of the sciences, and as such they do not belong in any science properly. On this count Galileo and Rocco agree. Galileo deployed this methodological stance in the *Dialogo*, that is, by assuming the motions of the earth around the sun and around its polar axis as first principles of cosmology he wanted to demonstrate that certain apparently inexplicable effects concerning the tides could be causally explained. His book was to have been entitled *Dialogue on the flux and reflux of the see* (to which the ecclesiastical censorship objected), in order to emphasize the physical argument from tides with which Galileo aimed at demonstrating the truth of the Copernican system. Rocco found Galileo's demonstration defective on commonly accepted methodological grounds. There was no

---

74 Spini 1983, pp. 161 ff. Rocco was a priest who taught philosophy in the convent of S. Giorgio Maggiore in Venice. He is believed to be the author of the *Alcibiade fanciullo a scuola*, a novel celebrating pedophilia, in which we also find, as in Vanini, a polemic against marriage and the praise of the philosophers who oppose it. Cf. Spini 1983, p. 165.

75 Cf. Galilei 1890–1909, VII, pp. 17 ff., for Favaro's comments on Rocco, and pp. 569–750, for Rocco's *Esercitationi* postilled by Galileo.

76 Galilei 1890–1909, VII, p. 577.

incommensurability between Rocco's and Galileo's theories such as has been fancied by modern philosophers and sociologists of science of positivist orientation. On top of this one has to keep in mind that both libertines knew how to duel with the same abrasive techniques of epideictic rhetoric. Both faced the same dilemmas that everybody faces who starts from the prejudice of first principles. They are indemonstrable within the accepted framework. There is no higher science in which first principles can be deduced from premises. They are open to investigation only by rhetorical strategies such as *ingenium*, the art of inventiveness, or metaphor and aesthetic inquiry, or alternatively they can be supported by empirical evidence as Galileo would show in his last book. Both Rocco and Galileo engage in the same movement away from the anonymity of the 'they' by retreating towards strange places, different ones, to be sure, that are forbidden by orthodox culture. Our difficulty in reading these heterodox libertines is that we have come to a point where we look backwards at them with the naïve prejudice that our theories are correct, and that theirs must be erroneous unless we find in them a precursor of modern views. Galileo's theories have always been regarded as such a precursor. But his argument from tides remained at best inconclusive and at worst flatly erroneous on his own principles, depending on how sympathetic your reading will be. Rocco's most damming objection was that Galileo systematically erred by applying the scheme of deduction that was called the 'error of the consequent'.

By marshalling astronomical evidence that seemed to be consistently deducible from the motions of the earth and the planets around the sun, Galileo inferred that the system of the universe described by Copernicus was true. Rocco noted that this logical reasoning was a fallacy such as that which would be incurred by somebody reasoning that since he feels hot, then fire must be close to him.[77] Galileo was taken aback. This fallacious inferential scheme was exactly how he had proceeded throughout in the *Dialogo* and especially in the argument from tides. By hypothesizing the Copernican motions of the earth, he deduced many details that are observed in tide phenomena. But then the whole argument was supposed to demonstrate that the tides prove the motions of the earth. It was the error of the consequent that Rocco had pointed out as the structural flaw in Galileo's book. Galileo's postil to this objection must have been a torment. He answered that Rocco did not know what he was talking about. He scolded him for his incomprehension. Surely there are many different causes of heat such as fire, friction, or the sun, insisted Galileo, and therefore it is quite correct that deducing fire from a sensation of heat is

77 Galilei 1890–1909, VII, p. 699.

erroneous. But the cosmological order of things is totally unique. Therefore, whoever looks for anything else than what is, he will look for the false and the impossible.[78] It is a riddle of an answer. Even granting to Galileo that there is one cosmos only, namely, the orderly arrangement of things in the universe, as he seems to posit in his premise, he would have to concede, I surmise, that the phenomena that we perceive with the senses might be explained by different orderly arrangements, or partial arrangements, within the one cosmos. He knew perfectly well that, for example, the motion of Venus around the sun that could be deduced from the newly observed phases of the planet could also be accommodated in the Tychonic system. Two incompatible arrangements of the universe were both compatible with the phenomenon of phases. He seems to be getting at something that transcends the inner structure of the cosmos by hinting at the totality of the order of things. There is one totality, he intimates, and therefore one way only of arranging the main bodies of the universe that is consistent to the fullest extent possible with observation.

It seems to me, however, that there is another interpretation of Galileo's postil that might be exhilarating. Let us concede to Rocco that he is quite right. Galileo's overarching argument in the *Dialogo* (especially the most prized argument from tides) is systematically flawed in terms of syllogistic logic, and he knew well that this was the case. But would it be really flawed in terms of libertine logic? The book is not a deductive sequel of syllogisms. Its structure is dialogic through and through. It is a dialogue among characters, sarcastically commented by the anonymous voice of the postiller. It is a game of epideictic logic. But, as I suggest, the libertine positions himself away from the 'they'. In this modality of being-in-the-world of libertinism even logic is seen as received wisdom. It is the 'logical-they' that must be called into question by the libertine's zeal for nudity. Logic covers up, like clothing. I hear Galileo, lo and behold Aristotelian libertines! Even though it cannot be undone, syllogistic logic can be suspended in the unfettered game of praise and blame. It can be shed like clothing in search for the erotic pleasures of nudity. Rocco was a hell of an unbeliever but he failed to appreciate the subtler play of nudity that requires undressing the clothes of syllogistic logic. Neither did Galileo appreciate his own game of brinkmanship in the

78 "Voi peccate per non intender quello che vi diciate. Sono in natura molti modi di scaldare, cioè col fuoco, col Sole, con la confricazione, col litame, con l'acqua e calcina, con la febre; e tutti questi sono in atto sempre: ma l'ordine del mondo è uno solo, nè mai è stato altrimenti: però chi cerca altro che quel solo che è, cerca il falso e l'impossibile" (Galilei 1890–1909, VII, pp. 699–700).

Dialogo any longer. When he answered Rocco he had already abjectly abjured in Rome, and seen his book burned in a blaze of glory. The game was over. But he had always known that it was a game. He tells us so. We only need to read the opening passage of the Fourth Day of the *Dialogo* in the light of Rocco's criticism. It is a disclaimer inaugurating the final showdown with syllogistic logic. "…let us demonstrate how nature has permitted that (regardless of whether the thing *in rei veritate* is really so, or rather it is a joke [intended by nature] to poke fun at our excogitations) the motions, with which the earth was endowed long ago in order precisely to satisfy the flux and reflux of the sea, are now found to be so peculiarly suitable to serve the cause of the tides, and [let us demonstrate] how vice versa the very flux and reflux comes to confirm the mobility of the earth".[79] The postiller comments: "La natura per ischerzo fa che il flusso e reflusso del mare applaude alla mobilità della Terra".[80] Nature jokingly makes the flux and reflux of the sea applaud to the mobility of the earth. It is a parlor game! Galileo is accurately describing his scheme of reasoning as 'confirmation', consciously avoiding the loaded language of logical demonstration or mathematical proof. But when his sensitivity is pricked by Rocco's criticism, he responds furiously, rummaging in his pockets for an explanation of the cogency of confirmation that, as I have suggested, does not seem to be very convincing. The Italian 'confermare' is modeled on the Latin 'confirmare', which, aside from the meaning of 'confirm' that passed into its Italian cognate, also has that of 'instill courage, inspirit, animate, embolden'. This latent meaning is still palpable in Galileo's libertine confirmation. Nature and Galileo-the-libertine applaud to the mobility of the earth. The flux and reflux of the sea emboldens Galileo to think that the Copernican system is true. But this is not all. Once the libertine mode of being-in-the-world is taken into account, Galileo's phrase '…il medesimo flusso e riflusso comparisca a confermare la terrestre mobilità' can be read as a veiled sarcastic reference to the sacrament of confirmation. Galileo's philosophical confirmation is structurally belonging in his being-in-the-world of libertinism. One can see the beginnings of the logical-positivist philosophy of

79 "…mostriamo, come la natura ha permesso (o sia che la cosa *in rei veritate* stia così, o pur per ischerzo e quasi per pigliarsi gioco de' nostri ghiribizzi), ha, dico, permesso, che i movimenti, per ogni altro rispetto che per sodisfare al flusso e reflusso del mare, attribuiti gran tempo fa alla Terra, si trovino ora tanto aggiustatamente servire alla causa di quello, e come vicendevolmente il medesimo flusso e riflusso comparisca a confermare la terrestre mobilità" (Galilei 1890–1909, VII, p. 442).

80 Galilei 1890–1909, VII, p. 442.

confirmation in Galileo's being-in-the-world of libertinism. The endurance of the puzzle of confirmation in scientific processes is the unrecognized legacy of the sacramental logic of early modern libertinism.[81] Salviati's recital of the flux and reflux is personified by Galileo as a religious minister that shows up at the ceremony to ritually impart thc sacrament of confirmation on the recipient, the novice-philosopher, in need of strengthening his faith in the mobility of the earth. There are sacraments that are necessary to salvation, and there are sacraments that operate to perfect salvation. Confirmation is of the latter species.[82] Before confirmation confession is required, though, so that the gift of the Holy Spirit can be received by the cleansed believer. In order to be perfect, namely, Christian, one must be anointed with the Episcopal chrism of confirmation.[83] The novice-philosopher has participated in the ritual of confession conducted by Salviati and Sagredo during the first three Days of the *Dialogo*. His mind has been cleansed of the sins of Aristotelian philosophy and of his egocentrism disguised as geocentrism. Standing on the threshold of the Fourth Day he is just about to receive confirmation, perfecting his Copernican faith, anointed with the chrism of the flux and reflux of the sea. Confirmation is not the ritualistic transmission of perfection from minister to novice but in fact an inner strengthening of faith that renders all recipients of the sacrament more perfect than they were before. Thus, it is a sacrament of self-transformation.[84] Confirmation as mode of being-in-the-world of libertinism has a transformative power.

Unlike syllogistic logic, which carries the truth value from premise to conclusion, the libertine logic of confirmation is a self-transforming ritual that perfects the philosopher in his faith. That Galileo avidly read, postilled and might have planned on answering Rocco's *Esercitationi* is quite extraordinary. The *Dialogo* was immediately attacked by two other philosophers of Aristotelian persuasion, Claude Bérigard and Scipione Chiaramonti.[85] Galileo

81 Cf. Carnap 1936, 1937, for the archetypal discussion of confirmation in logical-positivist philosophy of science. I will leave to the reader the pleasure of ruminating over the footnotes spawned by Carnap's masterful treatment over the last eighty years.

82 Lippomano 1553, p. 293 verso.

83 Lippomano 1553, p. 293 verso.

84 "Similmente la eccellentia a la quale si da questo sacramento, non s'intende per conto di un'huomo ad un'altro, ma come di uno medesimo a se stesso, si come quell'istesso essendo huomo perfetto, ha eccellentia rispetto a se, quando era fanciullo" (Lippomano 1553, p. 294 recto).

85 Bérigard 1632, Chiaramonti 1633. On Bérigard, see Ragnisco 1894, Favaro 1920, and Bellucci 1971. Chiaramonti's rebuttal would deserve serious scrutiny. Particularly, his reply

seems to have ignored them both, at least initially, and focused, energetically and exclusively, on Rocco. Why so? While the rebuttal by Chiaramonti was to be expected since in the *Dialogo* Galileo had demolished the former's theory of the sublunary origin of new stars, Bérigard might have deserved the same attention that he was devoting to Rocco. But if we glance at Bérigard's criticisms of the argument from tides, we find a plausible answer. Bérigard makes the same error in terms of syllogistic logic that Galileo makes. Naïvely unaware of the libertine game that Galileo was playing, and furthermore not having been trained in the Padua tradition of logical instrumentalism and of the philosophy of demonstrative science, Bérigard fell into the trap of answering Galileo's argument on its own logical terms, thus failing to flag the error of the consequent that Rocco had aimed his finger at. Bérigard pointed out that in the case of the annual periodicity of the tides, empirical evidence does not conform with Galileo's theoretical predictions, and that rather than the mobility of the earth it confirms its stability.[86] Bérigard was a honest worker but a poor logician. Had he concluded that empirical evidence disconfirms the earth's mobility he would have had some leverage. But he repeats the error of the consequent and draws the wrong conclusion that it *confirms* the stability of the earth. Galileo was too good a logician to waste time on Bérigard. Obviously he realized that Rocco's philosophical *Esercitationi* was of another order of magnitude.

In essence libertinism is play. It involves being-in-the-world in the pleasure of distancing itself from the 'they', though it still retains it within the

---

to the argument from tides is dramatically lucid. Galileo imagined that the composition of the daily and annual rotations of the earth would accelerate and decelerate periodically the waters of the world seas, basically every twenty-four hours. This periodicity of acceleration and deceleration was put forward by Galileo as the primary cause of the tides. Chiramonti destroyed the assumption that this composition of circular motions might cause either acceleration or deceleration. He imagined a bucket filled with water rotating regularly on a pivot while being moved uniformly forward along a smooth plane. He deduced that no acceleration or deceleration of the water would be caused by the uniform forward motion of the bucket. The argument is tantamount to an intuition of the impossibility that an observer can detect any effects within his reference framework of the rectilinear uniform motion at which he is moving, namely, of the principle of relativity, which also Galileo advanced in the *Dialogo*. It is conceivable that Chiaramonti retorted against him the principle that Galileo had illustrated with the celebrated thought-experiment of the ship sailing in uniform rectilinear motion. Cf. Chiaramonti 1633, pp. 334–335.

86 "Actum esset de terrae mobilitate, si ut ista perbelle sonant, ita inter se consentirent, nunc ita dissociata sunt, ut confirment magis terrae quitem, quam eam labefactent..." (Bérigard 1632, p. 52).

purview of the 'they' that define the prohibition zone for being-in-the-world to make its forays. The logic of confirmation that Galileo had deployed in the *Dialogo* showcased his daring libertinism. We should not downplay its significance. Like hereticality libertinism projects possibilities of understanding that cannot be otherwise disclosed. Being-in-the-world of libertinism entangles the individual and the anonymous others. It reveals that understanding the world is always understanding one's ownmost authentic possibilities of being. Galileo gradually shifted his focus from postil, through annotations, to creative writing. This progressive movement can easily be followed in the seventh volume of *National Edition*. It was this self-reflective movement that pushed Galileo away from the narrow project of a polemical answer to Rocco, that is, his initial abiding by the rules of the 'they' calling upon him to answer the Aristotelian point by point, as was customary. The postils turned into full blown creative writing. They mutated into the first concept of his final masterpiece, the *Discorsi e dimostrazioni matematiche intorno a due nuove scienze*, as they gave him the opportunity to reconstruct his intellectual autobiography and the reasoning processes underlying his discovery of the fundamental principle of the sameness of speed of all bodies freely falling in the void. What is even more interesting from the existential viewpoint is the mood transition throughout the *Postil to Rocco*. As he moves away from the libertine impulse that had been reawakened by Rocco, the mood of the writing changes dramatically. The vehemence of the postils, the virulence of the vernacular expletives, turns decisively towards the calmness of interiority. The vociferous space of libertinism gives way to a private cathedral of silence. Galileo seems to be entering a different dimension of the spirit. He ceases to address Rocco while imperceptibly begins to address himself. As the noise of polemics fades into distant echoes, self-reflectivity takes center stage, as if Galileo were in search of a more authentic revelation. There is no exaggeration in saying that what has been mundanely labeled 'Postils to Rocco', would have deserved the title of Galileo's eschatological writings. Here is a sample demonstrating the shift to the self-reflective tonality.

> …I formed an axiom such that nobody could ever object to. I hypothesized that any heavy body whatever which descends has in its motion degrees of speed so limited by nature and fixed, that it would be impossible to alter them, by increasing or decreasing its speed, without using violence in order to speed up, or slow down, its natural course. I then figured in my mind two bodies equal in volume and weight, such as, for instance, two bricks, which depart from the same height at the same time. These will doubtless descend with the same speed, assigned to them by nature. If this speed has

> to be increased by another mobile, it is necessary that this mobile move more swiftly. Yet if one imagines the two bricks joining together while descending, which one would be that by adding impetus to the other will double the latter's speed, given that speed cannot be increased by an arriving mobile if it does not move more swiftly?[87]

The mood of stillness that Galileo finds himself in after transitioning from noisy polemic to silence discloses the memories that help him retrieve himself from dispersion in the 'they'. He picks himself up in order to reconstruct his ownmost possibilities of being in understanding the principle of free fall that was called into question by Rocco. The void space upon which this understanding is projected is not first and foremost an abstraction, or idealization, in the sense of the purification of the ideal case of fall from the imperfections of concrete, material bodies tumbling downwards in the real world. Prior to this abstract thought of the void, traditionally rejected as impossible in Galileo's life-world, but in which the principle of fall will eventually have to be formulated, is the emptiness of interiority. The finding himself dejected alongside the libertine 'they' that was prompted by Rocco's sharp criticisms has changed into Galileo's facing his own silent self turning towards a call from nowhere. The void is nowhere in space. Falling away from the call is prior to imagining falling bodies in space. 'I then figured in my mind…', says Galileo. Imaginative figuration becomes active in the present because understanding seizes upon this possibility of being more authentically his own self than was the case in finding himself dejected alongside the 'they'. Once again it is not that the anonymous 'they' must be negated, for it is structurally a mode of being-in-the-world. Rather it is resisted in willingly facing the call from nowhere that makes the void a possibility of being in which understanding falls freely. This suffering of the will resists the violence of the abuse intrinsic in the logic of argument and counter-argument that became prominent in early modern Europe with the ascendance of the religious denominations. Both Protestant and Catholic reforms resorted to the extermination of the 'diverse' that was identified as 'unorthodox' thinking. Thought became the habeas corpus for conviction that was legitimately crafted during the inquisitorial procedure by means of physical and moral torture. The 'they' that is acknowledged by the libertine's act of defiance is the mobbing multitude, the philosophical *mobile vulgus*, that is at the same time everybody and nobody. Libertinism was never a social category, as Spini's ground-breaking study has demonstrated, but an

87 Galilei 1890–1909, VII, p. 731.

acknowledging finding oneself alongside the anonymous presence of the mob of philosophers.

Hermeticism

Spini has formulated the problem of Galileo's religion from an original viewpoint. Galileo was very reluctant to write about his personal convictions. The impression that the reader receives when perusing the massive correspondence collected by Favaro in the *National Edition* is disconcerting. Virtually nothing of the contemporary political and religious upheavals that Galileo witnessed transpires from the letters. Clearly Galileo was consciously and meticulously erasing every possible trace of language or opinions that might engender suspicion. Thus, his religious persuasions remain elusive because he did not want to write about religion, and above all because he did not want anybody to know 'publicly' about his personal religion. Letters could become easily public unbeknownst to the author, and on the other hand the author knew how to circulate his ideas if he so willed by simply writing a letter to the opportune correspondent. The sphere of private life as we know it nowadays did not exist in the early seventeenth century. Galileo's religion must be reconstructed from derivative evidence. Spini thinks that Galileo's God is similar to the God-Demiurge of Plato's *Timaeus*, but that Galileo's conception transcended Plato's ideas in that Galileo thought that God's operation had not been limited to creation but that God could intervene to remodel the universe after creation as many times as he saw fit. Spini's conjecture is based on Galileo's argument that God pushed the planets into their orbits by impressing on them a rectilinear motion of fall that was converted into circular motion when they had reached their correct distances from the sun.[88] I believe that this hypothesis is pointing us in the right direction. In the *Discorsi*, Galileo tells us something extremely important about his convictions. He says that after having calculated the heights from which the planets should have started their descent if their uniformly accelerated motion of fall was to deliver them right on target at their distances from the sun, and moreover after having satisfied himself that his calculations were empirically sound, he had nevertheless decided to

88 Spini 1996, pp. 29–31. Galileo's Platonic myth is recounted in slightly different forms in his two most important books, the *Dialogo* and the *Discorsi*. See Galilei 1890–1909, VII, pp. 43–44, 53–55, and VIII, pp. 283–284. Cf. Acerbi 2000, who has reconstructed the Platonic framework of ideas that might have inspired Galileo's myth.

refrain from talking about his findings to avoid that the novelties he had discovered, which had already provoked so much indignation, might cause even more rage.[89] Note that Galileo is emphasizing that he did not even want to *talk* about them, let alone write or publish anything. Though he diplomatically wishes to avoid more outrage, it is the intense longing for silence that we see once again looming large in his mind, striving to leave behind the clamors of disapprobation of the vulgar philosophers. Galileo's explanation of the miracle of Joshua is another instance of his God's interventionism that leads Spini to reach the following general conclusion. Galileo has an absolute repugnance to identify God with immobility.[90] According to Spini, Galileo's epistemology is consistent with this conclusion. Human science and divine science are not qualitatively different, they differ quantitatively because humans can only know some truths by laboring at them while God possesses the whole truth intuitively, and all truths are seen by him with equal clarity from his non-temporal perspective. Hence, Spini claims, the problem of Galileo's religion is the problem of deciding to what extent these disparate beliefs can be identified with Christianity. Galileo's philosophy is not so much a *theologia crucis* as a *theologia gloriae*.[91] Galileo's philosophy, in other words, does not place Christ at his center, but rather the burning desire to become equal to God that glories in the exercise of the human intellect.

I will proceed further along the lines of Spini's argument by making the suggestion that the missing element in the puzzle of Galileo's strenuous silence about his religious convictions is hermeticism. However, I reiterate that it is not the doctrinal contents of the hermetic writings (which circulated in the Renaissance in various formats and different editions) that I intend to focus on, and that I do not intend to make the suggestion that specific hermetic doctrines might have been the literary sources of Galileo's natural-philosophical theories (though this cannot be ruled out either). This philological way of approaching the problem of thought formation is committed to the positivist historiography that I wish to interrogate, and which, as I have noted, was already called into question by Croce. Rather, since my philosophical outlook is broadly phenomenological, I concentrate on the transformative

89 "*Salviati*. Mi par sovvenire che egli già mi dicesse, haver una volta fatto il computo, ed anco trovatolo assai acconciamente rispondere alle osservazioni, ma non haverne voluto parlare, giudicando che le troppe novità da lui scoperte, che lo sdegno di molti gli hanno provocato, non accendessero nuove scintille" (Galilei 1890–1909, VIII, p. 284).

90 "C'è solo una ripugnanza assoluta a identificare Dio con l'immobilità" (Spini 1996, p. 30).

91 Spini 1996, p. 31.

moods that might have been communicated through the hermetic writings, which have a performative structure that invites the reader to take part in strange processes of self-transformation, and will ask questions about their efficacy in Galileo's life-world. I look at hermeticism as a modality of being-in-the-world that projects possibilities of understanding and imagination as moods and tonalities, thus setting, so to speak, the scene of the learning process through an attunement that is prior to, or at least concomitant with, the infusion of doctrinal contents. Particularly, I will consider some excerpts from the set of writings that were translated from Greek into Latin by Marsilio Ficino and which must be considered as being part and parcel with the Florentine Renaissance culture that Galileo was imbued with. One should also remember that textual reconstruction has its own intrinsic limitations due to the fact that oral styles of cultural transmission were still highly influential in early modern educational and intellectual milieus. Galileo is exemplary in this regard. His writings are replete with citations from Aristotle that are presented as verbatim quotes, typically introduced by phrases such as 'these are Aristotle's words…', and yet we can be fairly sure that they were written down from memory since often the words do not correspond exactly to the texts that can be retrieved from any known editions. Because of the value of its poetic prose I prefer the already mentioned Italian edition of Marsilio Ficino's *Poimander* (Il Pimandro) that was translated by Tommaso Benci, and which bears witness to the popularization of hermeticism in the vernacular culture well beyond the learned circles.

Positivist textual reconstruction was the philological model that Francis Yates applied to her masterful study of Renaissance hermeticism, *Giordano Bruno and the hermetic tradition*, which, in her intention, simply aimed at placing Bruno in the hermetic tradition.[92] However, notwithstanding the wonderful erudition that Yates handles with great acumen, her method is not beyond question in many respects. The unstated assumption behind the philological model seems to have been that a similarity of doctrinal contents, more or less consciously perceived by the author, would imply a hidden process of transmission of thought or ideas or entire doctrines from the source to the reader. While it is of course perfectly reasonable that a reader might absorb and/or reject certain elements from her reading materials, the model pushes the author to repress the consciousness of her own involvement in the hermeneutic process. This becomes apparent when one scrutinizes the comparative analyses of texts that Yates furnishes as objective philological

92 Yates 2002 (1st ed. 1964), p. xii.

proofs. I would like to focus on one example discussed by Yates because it is instructive to compare and contrast the strengths and weaknesses of both the positivist model and the phenomenological hermeneutics that I practice. She argues that a recurrent hermetic theme in Bruno is the glorification of Egyptian magic. She quotes a long passage from Bruno's *Spaccio della bestia trionfante*, and draws the conclusion that Bruno's argument is "[n]eedless to say [...] based on the god-making passage in the *Asclepius*", which she has summarized at the beginning of her book.[93] However, I find it very difficult to see what Bruno might have borrowed from the *Asclepius* passage in view of the claims he is making in this section of the *Spaccio*.[94] In fact, to my mind, Bruno is saying something quite different from the *Asclepius*, and quite possibly against the *Asclepius* magical doctrine of god-making.[95] Briefly, while Asclepius advocates the magical invocation of daemons to enliven the statues of the gods made by humans so as to turn them into man-made gods, Bruno says that living things represent gods and that God is in all things, and moreover that

93 Yates 2002 (1st ed. 1964), p. 234, for the quotation, and p. 39, for the summary from the *Asclepius*. I read the Asclepius passages referred to by Yates in Ramelli 2005, pp. 556–559, 582–587, which is based on the same text read by Yates (namely, Nock and Festugière 1945/1954).

94 "'Però' disse, 'diverse cose vive rapresentano diversi numi e diversi potestadi: che oltre l'essere absoluto che hanno, ottegnono l'essere comunicato a tutte le cose secondo la sua capacità e misura. Onde Idio tutto (benché non totalmente, ma in altre più e meno eccelentemente) è in tutte le cose. Però Marte si trova più efficacemente in natural vestigio e modo di sustanza non solo in una vipera e scorpione, ma et in una cipolla et aglio, che in qualsivoglia maniera di pittura o statua inanimata. Cossì pensa del Sole nel croco, nel narciso, nel' elitropio, nel gallo, nel leone: cossì pensar devi di ciscuno de gli Dei per ciascuna de le specie sotto diversi geni de lo ente; perché sicome la divinità descende in certo modo per quanto che si comunica alla natura, cossì alla divinità s'ascende per la natura, cossì per la vita rilucente nelle cose naturali si monta alla vita che soprasiede a quelle'. 'È vero quel che dici,' rispose Momo, 'perché in fatto vedo come que' sapienti con questi mezzi erano potenti a farsi familiari, affabili e domestici gli dèi che per voci che mandavano da le statue gli donavano consegli, dottrine, divinazioni, et instituzioni sopraumane: onde con magici e divini riti per la medesima scala di natura salevano a l'alto della divinità, per la quale la divinità descende sino alle cose minime per la comunicazione di se stessa...'" (Bruno 2002, II, pp. 354–355).

95 "....quoniam ergo proavi nostri multum errabant circa decorum rationem increduli et non animadvertentes at cultum religionemque divinam, invenerunt artem qua efficerent deos. Cui inventae adiunxerunt virtutem de mundi natura convenientem eamque miscentes, quoniam animas facere non poterant, evocantes animas daemonum vel angelorum eas indiderunt imaginibus sanctis divinisque mysteriis, per quas idola et bene faciendi et male vires habere potuissent" (Ramelli 2005, pp. 582–584).

as God descends into nature so it is possible to ascend from nature to God. It is the gods who speak to humans from the statues but there is no suggestion that humans are god-makers. Thus, I would argue that the hermetic doctrines of god-making put forward in the *Asclepius* were not imported as raw materials by Bruno into his *Spaccio*. If there was a process of tradition building, it cannot be explained by a similarity or dissimilarity of doctrinal content. Since it is not credible that Yates misread Bruno's Italian prose, as she was unquestionably competent in the European languages of the Renaissance, I suggest that the philological model obscured consciousness of her involvements with the literary material she was interpreting. In my perspective, she projected possibilities of understanding on the historical canvass of hermeticism that she was unfolding, and which were touching and transforming her being-in-the-world, yet the philological model of objectivity prevented her from resisting the call from the anonymous 'they' of positivist scholarly expectations.[96] However, I do not regard my phenomenological viewpoint as privileged (indeed I wish I knew more about Yates's being-in-the-world as she could perceive such strong similarities between Bruno's *Spaccio* and the *Asclepius*), and I revere the comparative analysis of textual materials on sound philological basis as absolutely indispensable. But one virtue of the imaginative hermeneutics I discussed in the *Preface* is vigilance about one's involvements, and the incitement to respect and accept the emergence of unexpected phenomena in consciousness that might not appear relevant at first sight, and to make an effort to lay them out for scrutiny, courageously, in order to integrate them in the processes of inquiry, appropriation, and self-transformation.

Fascinating examples of performative texts from the *Corpus Hermeticum* abound.[97] In the fifth discourse, entitled 'God is both hidden and manifest', Hermes speaks to his son Tat as follows.

---

96 One has only to read in between the lines of her preface to see how her realization that Hermeticism was the key to understanding Bruno, a true illumination, changed her outlook profoundly. I felt this moment of relief myself as I began to take seriously the clues to Galileo's hidden life-world.

97 The presence of the *Corpus Hermeticum* in Renaissance culture has by and large been studied in view of its philosophical doctrines, and the work by Yates exemplifies the tendency to look for doctrinal precedents. More recently, however, it has been recognized that the *Corpus Hermeticum* originally carried religious and ritualistic meanings that cannot be easily dissociated from the philosophical arguments. "Yet in most cases the arguments quickly tend to get formalized and pass easily into a much more performative discourse that at its peak resembles the language of ritual more than that of philosophy. [...] It is typical of Hermetic arguments that while based in Greek philosophy, they often disregard

My Tat, I will consider this discourse, aside from what has been said, so that you will not miss the most important name of God, and furthermore so that you will know that it is absolutely manifest what others think is occult. For, if he is not manifest somewhere then he must be nothing. That which is manifest to us is generated, but that which is hidden is everlasting. For, it is not necessary that he makes himself manifest in that he is never deprived of being. Certainly, he lays open before our eyes all things, and yet he hides himself, that is, he who is endowed with everlasting life hides himself in obscure places while he brings to light everything. Nevertheless, he demonstrates all things to the imagination. For, imagination captures only things that are generated, and nothing in the imagination escapes generation. But the un-generated one is not comprehensible to the imagination. Through him all things are manifest and he shines through all things. He becomes manifest especially to those with whom he has chosen to communicate. You, my son Tat, pray the lord, Father, the one from whom the one is, so as to become worthy of his mercy. Thus you will comprehend God, if one of his rays will illuminate your intellect benignly. For, only hidden understanding sees hidden things. Therefore, if you, Tat, will regard him with the mind's eyes, he will manifest himself to you. Certainly God shines everywhere and in all places of the world without envy, and he is so manifest that not only can we understand him but we can, so to speak, touch him. For, everywhere he is before our eyes, and his image is visible. But if your inner light is hidden from you, how can you possibly find yourself or him? Indeed, my son Tat, when you want to see God, look at the Sun on high, think of the course of the Moon, consider the order of all the Stars, and how all of them preserve their order. [...] Might God be willing, my son, to give you the strength to fly up to the highest region of air with wings, reaching the middle region, between Earth and the heavens, so as to regard the solidity of the Earth, the vastness of the Seas, the flow of Rivers, the breath of Air, the life and quickness of Fire, the course of the Planets, and the rapacious swiftness of the heavens! [...] What an exquisite vision, that you might see with one movement of the eyes the whole cosmos, and the stasis of the immobile maker, whom you might see even though he is hidden.[98]

---

its logical strictness and emphasize the rhetorical and dramatic aspect instead". Cf. Chlup 2007, pp. 144–145.

98 "O Tazio, io tratteró, oltre a quello che é detto, questo parlare; acció che non ti manchi il precipuo nome di Dio: et che ti sia noto, essere molto manifesto, quello che a molti pare occulto. Imperó che, se egli non si manifesta in luogo alcuno, certo e' sará niente. Quello, che si palesa al nostro aspetto, è generato: ma quello, che si nasconde, è sempiterno. Imperó che non è necessario, ch'egli apparisca: imperó che mai non manca d'essere. Certo esso pone avanti agli occhi, tutte l'altre cose; et esso sta segreto: cio è, quello, il quale usa la sempiterna vita, mentre che chiaramente a luce conduce ogni cosa, nelli luoghi occuti si nasconde: Nondimeno elli dimostra ciascuna cosa, alla fantasia. Imperó che, la fantasia s'adopera solo, circa quelle cose, che sono generate: nella quale è niente fuori della generazione. Ma l'unico et non generato, è incomprensibile alla fantasia; ma conció sia che per lui tutte le cose sieno manifeste, ancora in tutte, et per tutte le cose risplende. Et spezialmente si palesa, a quelli, a' quali egli ha voluto communicare la notizia di

The performative discourse steadily builds up the emotional tension that is eventually relaxed in the augural invocation to God that he might grant Tat the force to fly upwards with wings. The crescendo of the fragment transports the imagination towards the middle region of the universe where Tat can contemplate the fabric of the cosmos. Eventually, the imagination is released from the order of generation in which it appears to be initially constrained, and thus Tat will be able to see the hidden maker of all things. It is an astonishing thought-experiment that one is invited to perform by launching the imagination through space. It challenges the reader to turn inwards to inspect the inner light that might be hidden from consciousness. God is always shining through the material universe yet hiding at the same time. But the force of the imagination seems to elevate the intellect towards a 'place' where the manifestation of the everlasting order of the cosmos brings about the revelation of the hidden God-maker. The performance of the exercise creates a mood of cosmic vastness that dilates the space of the reasoning intellect. More possibilities of understanding the presence/absence of the God-maker throughout the cosmos are thus disclosed by this mood than the reasoning intellect could have attained by deductive logic. The doctrinal content of the piece recedes in the background. Not that it is irrelevant, for its severance would disinte-

---

se. Tu adunque, o figliuolo mio Tazio, in prima, con pietosi prieghi, priega il signore Padre unico, et da cui è l'uno, che tu sia degno della sua misericordia. Et così, in tanto finalmente potrai intendere Dio: Se pure solamente uno de suoi razzi, benignamente illuminerá, la tua intelligenza. Certamente, solo lo nascoso intendimento, vede le cose nascose. Adunque, se tu lo riguarderai con gli occhi della mente, credimi, o Tazio, che egli ti si manifesterá. Certo, Dio, senza alcuna invidia, per tutte le parti del Mondo, in ciscuno luogo, risplende: et in tanto si manifesta, che non solamente intendere, ma ancora per un modo di parlare ci è quasi lecito, toccarlo con le mani. Imperó che, da ogni parte a' nostri occhi si mette innanzi, et manifesta: et vedesi la sua imagine. Ma se il tuo interiore lume, ti si nasconde: in che modo te o lui troverai? Finalmente, o figliuolo mio Tazio, quando tu pure vorrai vedere Dio: Riguarda su il Sole, poni mente il corso della Luna, considera l'ordine di tutte quante le altre stelle: et come ciascuno di loro conserva il perpetuo ordine. [...] Or' volesse Dio, o figliuolo, che ti fusse conceduto la possanza, che con l'aiuto delle alie, tu volassi nella piú alta parte dell' Aria: et posto nella mezzana regione, tra il Cielo et la Terra, ragguardassi la soliditá della Terra, la diffusione del Mare, il corrimento de' Fiumi, la larghezza dell' Aria, la vita e la celeritá del fuoco, il corso de' Pianeti, et la rapace velocitá del Cielo: o fligliuolo, o cosa felicissima a vedere: o beatissima visione, che certo con uno movimento d'occhi l'ordine di tutto il Mondo comprenderesti: et il fattore immobile per alcuno modo, mosso, il quale benche nascosto sia, ancora chiaramente vedresti" (Trimegisto 1549, pp. 33–36). (I have slightly modernized punctuation and occasionally the inconsistent orthography).

grate the unity of the performance. But the mood discloses the possibility of understanding one's cosmic positioning vis-à-vis the order of the universe and the God-maker. The mood of vastness expands the horizon of visibility that circumscribes one's finding oneself alongside nature in a state of dejection in which one does not understand. Moods dilate the space of reasoning by disclosing unsuspected possibilities of understanding.

Galileo had an obscure obsession with horizontality and verticality. He has his alter-ego Simplicio doubt that the demonstration of the parabolic trajectory of projectiles, based on the composition of horizontal, uniform, rectilinear motion and vertically accelerated fall on the surface of the earth, fails on account of the circular profile of the earth. There is no rectilinear horizontality, truly, charges Simplicio, it is all mathematical fiction. What will the value be of a demonstration of the parabolic trajectory of projectiles that assumes the false premise of the existence of rectilinear horizontality on the earth? Galileo's answer is astonishing. It must be taken phenomenologically at face value. First, Galileo states that in the demonstration of the quadrature of the parabola Archimedes assumes that all vertical lines from the points of a horizontal balance can be taken to be parallel, even though in reality they converge to the center of the earth. It is a false assumption. Yet the proof is valid. What we really perform, according to Galileo, when we follow Archimedes's 'contemplation' (*contemplazione*), is a thought-experiment in which we imagine being transported at an infinite distance from the center of the earth. If the distance is infinite, surely the vertical lines from the balance to the center of the earth will be parallel. Galileo does not even mention the balance being transported at an infinite distance, he simply imagines that we are transported away from the earth.[99] The reasoning that clinches the proof of the parabolic trajectory is embedded in a *contemplazione* made possible by a mood that dilates the space of the imagination. The geometrical argument cannot be dissociated from the *contemplazione*. It is the mood that alleviates the paradox. In Galileo's cosmological thinking the lines converging to the center of the earth are the directions of the weights hanging from the balance. But it is the center of the earth that determines the nature of gravity. The center cannot be imagined away if weight is to operate, for weight is the tendency of heavy bodies to fall towards the center. It is we who must imagine being taken to an infinite distance from the center so as to attain the

99 Galilei 1890–1909, VIII, pp. 274–275.

contemplazione that makes the mathematical proof go through.[100] The mood of Galileo's proof of the parabolic trajectory of projectiles is in the same tonality as the hermetic mood of vastness. Simplicio finds himself in a state of dejection in which he does not understand. Galileo invites him and the reader to exercise the imagination, the *contemplazione* that will change their mood from dejection to understanding. The mood of vastness is the dual of the mood of dejection. In the eleventh *Discourse*, mind says to Mercurius. "But if you submerge your soul in your body, and if you make nothing of it, blemishing it with words such as these, 'I know nothing, I cannot know, I sink into the horrible depths of the sea, I cannot fly to the heavens, I do know what I am now, or what I will be in the future'. Now, then, what will you have to do with God?".[101] The *Corpus Hermeticum* is continually awakening us to the possibilities of understanding, or vice versa alerting us to its failures. Moods are primordial presences constituting our being-in-the-world that the *Corpus Hermeticum* emphasizes. The preoccupation with moods betrays an implicit notion of truth which challenges the scholastic notion of adequation of intellect and thing by shifting the focus of adequation from intellect vs. thing to being oneself vs. being God. Truth means becoming God. "If you do not make yourself equal to God, you will never understand God".[102] This making oneself equal to God is a transformation, an imperative that involves being-in-the-world in its unity.

Strictly connected with the attentiveness to moods, we find in the *Corpus Hermeticum* a notion of the continuous activity of God as intrinsic in his essence. God himself appears to be moody, and apprehension of this God-maker requires the correct training of one's being in terms of proper moods. In my view, Spini is possibly correct in surmising that Galileo has a repugnance to identify God with immobility, and I think that, if this is the case, the *Corpus Hermeticum* might supply one missing link in Galileo's *theologia gloriae*. In the fifth *Discourse*, we are told that "it is impossible that God always be if He does not always make all things, that is, in Heaven, Air, Earth, Sea, and in the whole World, and in each single part of the World, both in that which is, and in that which is not".[103] In the eleventh *Discourse*, again, we

100 "Aggiungo qui, che noi possiamo dire che Archimede e gli altri supposero nelle loro contemplazioni, esser costituiti per infinita lontananza remoti dal centro..." (Galilei 1890–1909, VIII, p. 275).

101 Trimegisto 1549, p. 87.

102 "Se tu non ti adequerai a Dio; giá mai Dio non intenderai" (Trimegisto 1549, p. 86).

103 Trimegisto 1549, p. 38.

learn that "the power of God is eternity, and the work of eternity is the world, made not once, but always made since eternity".[104] The exercise of the active imagination that arrives at the vision of the God-maker is itself an exercise in the work of creation. The entire *Corpus* is suffused with the spirit of praxis, presented as the royal road to making oneself authentically equal to God. In the light of the *Corpus Hermeticum* Galileo's *theologia gloriae* appears to be a little less enigmatic.

Hermetic Science

The *Discorsi e dimostrazioni matematiche intorno a due nuove scienze attinenti alla mecanica ed i movimenti locali* (or *Two new sciences*, the English title) is a mysterious book ruptured throughout with symptoms of silence. Like the *Dialogo* the *Discorsi* is divided in four sections, called Days, during which conversations are held among friends. The Third and Fourth, on naturally accelerated motion and on the parabolic trajectory of projectiles, hint at the human being-in-the-world becoming equal to the God-maker by the imagination of knowledge. In them the three famous characters, Simplicio, Sagredo, and Salviati, read a book, from which theorems and demonstrations written in Latin by an Academician (Galileo) are read out aloud. But Simplicio, Sagredo, and Salviati are Galileo's masks. They intertwine discourses in the vernacular with the Academician's mathematical sentences. Magicians were often portrayed as reading from a book in order to perform supernatural transformations. In the *Discorsi* a hermetic science emerges that overcomes the libertine logic of the *Dialogo*. While the latter was contrived as confirmation, Galileo's new science of motion deploys the art of *ingenium*, or hermetic imagination, in order to elaborate a discourse on first principles that will jumpstart the science. It is the silences and keywords sparsely disseminated in the *Discorsi* that point to hermetic tonalities, which in turn invite us to draw attention to Galileo's personal transformation. His being-in-the-world has resulted in a marked attenuation of the libertine play of violence sustaining the *Postils to Rocco*. This becomes evident in regard especially to the mask of Simplicio. He is no longer the stubborn underdog of the *Dialogo*. He is Galileo's younger self. In the *Discorsi* Galileo turns to interiority in search of an inner light. We might say that the *Discorsi* is a hermetic autobiography whose silences are as important as the speeches. The First Day is weaved around the theories that Galileo held in his

104 Trimegisto 1549, p. 75.

youth about falling bodies in fluid media. This is supported by our knowledge of the early manuscripts on motion that were never published by Galileo in their original format but became the basis of the First Day five decade later.[105] The entire material that Galileo recycled in the *Discorsi* had been elaborated years before with the possible exception of the calculations of the ballistic tables associated with the theory of the parabolic trajectory. The earliest draft of the book's concept is the *Postil to Rocco*, in which Galileo began to shape his approach. This autobiographical reconstruction cannot be dissociated from the *nuove scienze* that he advances in the book.

Perhaps the most dramatic of Galileo's silences concerns Theorem 6, which states that the times of descent from a point to all points that are on a circumference whose diameter is perpendicular to the horizon are equal. The result is so startling that the interlocutors stop to think about it for a while. If bodies start moving from a point at the same instant, from rest, along all possible inclined lines in space, going in all directions, then at each subsequent instant of time all bodies will be on the same, expanding spherical surface. In other words, all bodies will be located at all times on an ever expanding spherical surface. Sagredo notes that according to this theorem expanding circles or spheres are generated from the kind of motion which nature itself makes use of, namely, naturally accelerated motion. He talks of 'emanation' from the point of origin of all motions. Then, Simplicio intervenes recognizing the cosmogonic character of the theorem. He comments that "it is quite possible that a certain great mystery is hidden in these true and admirable conclusions; a mystery, I say, concerning the creation of the universe, which is believed to be of a spherical shape, and concerning the abode of the first cause".[106] To which, Salviati replies that he totally agrees, even though, he notices, such profound contemplations are more the preserve of higher doctrines than the subject matter of the *Discorsi*. We, Salviati adds cautiously, are like those humble quarry-workers who extract the blocks of marble from which great sculptors can produce marvelous sculptures.[107] In reference to our knowledge of God, the *Corpus Hermeticum* says that "this knowing is divine silence, and a fixed concentration of all senses".[108] We have already seen the hermetic demonstration of the parabolic trajectory. Here I wish to discuss the

105 See Galilei 1890–1909, I, pp. 243 ff., the early manuscripts known as *De motu*.

106 Galilei 1890–1909, VIII, p. 225.

107 Galilei 1890–1909, VIII, pp. 225–226.

108 "La cognizione di quello, è divino silentio, et una intenta applicazione di tutti i sensi" (Trimegisto 1549, p. 61).

discourse by which Galileo shifts the focus from confirmation to the question of first principles.

The structure of the new science of uniformly accelerated motion is fairly simple. It aims at presenting a demonstration of the proportionality of the spaces traversed to the squares of the elapsed times in uniformly accelerated motion. This proof is achieved in Theorem II. The remaining theorems of the Third Days depend on this fundamental result. Although Galileo had known this result since his early researches in Padua, it has always been a puzzle why he procrastinated publication for so long. I think that one important reason might be that he had not been able to formulate a satisfactory scheme of deduction within which the theorem could be proved. This situation lasted until Rocco's book prompted him to turn inwards and question the libertine logic of confirmation. The logical structure of the new science can be summarized with the following quotations from the *Discorsi*.

> *Definition*. I call equably, or uniformly accelerated motion that, which starting from rest, adds to itself equal moments of speed in equal times.
> *Principle*. I assume that the degrees of speed acquired by the same mobile on differently inclined planes are equal when the elevations of the planes are equal.
> *Theorem* I. The time with which a space is traversed by a mobile, in uniformly accelerated motion from rest, is equal to the time in which the same space would be traversed by the same mobile in uniform motion, the degree of speed of which is one half of the maximum and final degree of speed of the previous uniformly accelerated motion.
> *Theorem II* (also known as times-squared law of fall). If a mobile descends from rest in uniformly accelerated motion, the spaces traversed by the mobile in any times are to one another in the duplicate ratio of the times, namely, as the squares of the times.
> *Comment*. What has been demonstrated in motions occurring along perpendiculars, must be understood to occur also in the same way in planes of any inclinations; for, in these we assume that the degree of acceleration increases in the same ratio, i.e., according to the increment of time, or, in other words, according to the simple and primary series of integers.
> *Theorem III*. If on an inclined plane and on a perpendicular, the elevations of which are the same, a mobile moves from rest, the times of the motions will be to each other as the lengths of the plane and of the perpendicular.[109]

The scheme hinges on the fundamental principle that Galileo assumes as a postulate. Since he has no proof of this principle he must look for a strategy to justify its validity. It is to be noted that this approach to the justification of a first principle is highly original. In Euclidean geometry postulates are given

109 Galilei 1890–1909, VIII, pp. 197–217.

as intrinsically knowable and no justification is presented. Theorems I, II are independent of the principle. Theorem III underpins Theorems IV, V, VI, the pillars of the new science of motion, which support its extension to uniform accelerations along planes of different inclinations. Thus, for the new science to be grounded a discourse must be invented to justify the principle. Galileo's evidence for the principle rests wholly on experience. Imagine a wall vertical with respect to the horizon, and a lead ball of 1–2 ounces, suspended by a string, 2–3 braccia long, perpendicular to the horizon, and hanging from a peg nailed in the wall. Draw a horizontal line on the wall, at a right angle with the perpendicular. Remove the pendulum to the height of the horizontal line. Let it go. You will see it descend, travel beyond the lowest point, and rise almost to the level from which it was let go. Were it not for the impediments of the air and of the string, the ball would arrive exactly at that level. From this Galileo concludes that the impetus acquired by the ball, descending to the lowest point, is enough to push the ball to the same height from which it came. Now imagine B being the degree of speed at the lowest point of the circular trajectory of the pendulum's bob, and imagine nailing a second peg on the wall, below the first one. We can change the position of the peg, which stops the string of the pendulum at some point below the pivot on the vertical. The position of the second peg determines the inclination, i.e., the radius, of the circular path of the bob after the string twists around the second peg. Observations can be made under similar circumstances by positioning the second peg at different heights. Through direct observation of the bob's swinging to a second point as high as its starting point, after the string twists around the stopping peg, a correlation can be determined between (a) degrees of speed acquired after fall from rest along circular paths having the same elevations but different inclinations (i.e., different radii of the arcs) and (b) the inclinations. The correlation justifies the principle if and only if *an analogy is made between the circular trajectories of the swinging pendulum and the rectilinear chords subtended by the circular arcs* that represent the differently inclined planes referred to by the principle.[110] It is the analogy that, as we shall see presently, requires the exercise of the imagination. For, while the arcs of the swinging bob are continuous curves, the chords are rectilinear segments that form an angle at the bottom like a v-shape. Following the demonstration of Theorem II Galileo presents the discussion of the celebrated incline plane experiment

110 Galilei 1890–1909, VIII, pp. 205–208. I easily repeated the experiment in a classroom with the help of undergraduate students. There was general satisfaction that Galileo's claim that the bob swings up to the same height as the starting point is acceptable.

with bronze balls rolling down a wooden beam at different inclinations. What is the role of this famous experiment in the scheme of the new science? This is the crucial question. Galileo's answer shows his rethinking of the libertine logic of confirmation. Initially, Salviati speaks rather loosely about confirmation of principles by experience in response to Simplicio's doubts.[111] But what he means is that the inclined plane experiment is supposed to give some empirical evidence for the times-squared law of fall. It is not supposed to confirm the principle of the same speed along differently inclined planes that has been assumed as a postulate, for this would immediately entail the error of the consequent that plagues libertine logic. Galileo guardedly comments: "Let us, then, for the time being, take this [the principle] as a postulate, the absolute truth of which will be established by observing that many other conclusions built on this hypothesis will respond to and compare precisely with experience".[112] Libertine confirmation is ruled out in its crude form. The novel logical scheme suggests that it is the entire fabric of the science of accelerated motion built on this hypothesis that will establish the truth of the principle. In other words, the burden of demonstrating the truth of the principle is shifted and spread out throughout the whole empirical basis of the science. So far so good, one might congratulate Galileo.

But the drama of the imagination is the analogy that is required for the justification of the principle, without which the whole logical fabric of the new science caves in. Here lies the crux of the matter, which Galileo calls the 'intoppo'. It is the already mentioned v-shape (something like the obstacle due to the angle) formed by the chords that are supposed to be the analogues of the circular arcs along which the bob swings back and forth.[113] When the ball reaches the bottom of the plane, impacting against the *intoppo*, will its course still be smoothly deviated upwards, thus imitating the swinging bob? By removing in the imagination the obstacle, one might see the ball

---

111 Galilei 1890–1909, VIII, p. 212.

112 "Prendiamo dunque per ora questo come postulato, la verità assoluta del quale ci verrà poi stabilita dal vedere altre conclusioni, fabbricate sopra tale ipotesi, rispondere e puntualmente confrontarsi con l'esperienza" (Galilei 1890–1909, VIII, p. 208).

113 "…anzi è credibile che, formandosi angoli da essi piani retti nel termine B [i.e., the bottom point], la palla scesa per l'inclinato secondo la corda CB, trovando intoppo ne i piani ascendenti secondo le corde BD, BG, BI, nell'urtare in essi perderebbe del suo impeto, né potrebbe, salendo, condursi all'altezza della linea CD: ma levato l'intoppo, che progiudica all'esperienza, mi par bene che l'intelletto resti capace, che l'impeto (che in effetto piglia vigore dalla quantità della scesa) sarebbe potente a ricondurre il mobile alla medesima altezza" (Galilei 1890–1909, VIII, p. 207).

moving upwards and hopefully doing what it is supposed to do for the analogy to work. But imagining away the *intoppo* is no mean task. How can you imagine removing the v-shape causing the *intoppo* without destroying the shape itself? Here we must insist on exploring the life-world for clues about this *intoppo*. Throughout the *Discorsi* Galileo is keen to separate what he thinks is knowable by science from what is not. What is often not knowable is certain accidents such as fluid resistance to the motion of projectiles caused by impediments. In scholastic philosophy the accident is knowable and separate from substance (generally not knowable, at least directly), while Galileo seems to be telling us that very often even accidents are not knowable because they vary in infinitely unpredictable ways.[114] Hence the question arises of the type of accident or impediment that the *intoppo* opposes to the motion of the rolling balls. Is it possible to abstract from the *intoppo* in the imagination?

Vincenzo Viviani wrote the first and most famous biography of Galileo. Viviani tells us that Galileo had a prodigious memory, and that he had memorized many works of classical Latin poets, and almost the whole epic poem by the celebrated Renaissance poet Ludovico Ariosto. To the admirers who praised Galileo's clarity of thought, Viviani recounts, Galileo used to answer that he owed it solely to Ariosto.[115] Indeed, those who familiarize themselves with both Ariosto's long poem and Galileo's dialogues will feel that Viviani's account is truthful. Galileo's absorption of Ariosto's poem modified Galileo's existence. Learning and memorizing leads to subtle reconfigurations in one's existential situation. Imagine the effects on being-in-the-world that the process of memorizing a huge epic poem such as Ariosto's will produce. The dejection in which one finds oneself cast away, the estrangement that does not understand, that is despondent and that does not know the whence and whither of this despondence, is a tremendous restriction of the scope of cognition. What is interesting in Viviani's account is that the modification induced by that feat of memory had a lasting cognitive effect which went beyond the mere retrieval of data from memory. That effect was described by Galileo and his admirers as *clarity* of thought. Human imagination can become like the imagination of the hermetic God, for His imagination is pure vision with

---

114 "De i quali accidenti di gravità, di velocità, ed anco di figura, come variabili in modi infiniti, non si può dar ferma scienza: e però, per poter scientificamente trattar cotal materia, bisogna astrar da essi, e ritrovate e dimostrate le conclusioni astratte da gl'impedimenti, servircene, nel praticarle, con quelle limitazioni che l'esperienza ci verrà insegnando" (Galilei 1890–1909, VIII, p. 276).

115 Galilei 1890–1909, XIX, p. 627.

equal clarity from a non-temporal perspective. Poetry seems to be able to induce a clarity of imagination in the human world that has the potential to become like God's imagination. Whoever has fresh in their mind the first few Cantos of Ariosto's poem, cannot fail but hear, when chancing upon Galileo's worry about *intoppo*, the poetic phrase *duro intoppo* resonating from one of the most intriguing octaves of the poem (Octave 56, Canto VI). But we can form an idea of how the hermetic imagination can be trained in the Ariostean fashion by memorizing the octaves in which the poet describes the *duro intoppo* (Octaves 61, 62, 63).

Canto VI (Edited by Cesare Segre)[116]	Canto VI (Trans. William Stewart Rose, with a slight modification in Octave 63)[117]
56 'Ma che non pensi già che seguir possa il suo camin per quella strada troppo: incontro avrà di gente ardita, grossa e fiera compagnia, con duro intoppo. Alcina ve li tien per muro e fossa a chi volesse uscir fuor del suo groppo.' Ruggier quel mirto ringraziò del tutto, poi da lui si partí dotto et instrutto.	56 'But that he must not think he shall pursue The intended journey far; since by the way He will encounter with a frequent crew, And fierce, who serve as rampart to the fay, That block the road against the stranger, who Would break her bounds, and the deserter stay.' Rogero thanked the tree for all, and taught, Departed thence with full instructions fraught.
61 Non fu veduta mai più strana torma, più monstruosi volti e peggio fatti: alcun' dal collo in giú d'uomini han forma, col viso altri di simie, altri di gatti; stampano alcun con piè caprigni l'orma; alcuni son centauri agili et atti;	61 Was never yet beheld a stranger band, Of mien more hideous, or more monstrous shape. Formed downwards from neck like men, he scanned Some with the head of cat, and some of ape;

116 Ariosto 1999, Electronic Edition, freely retrievable online at: http://www.liberliber.it/
117 Ariosto 1858, I, pp. 94–95.

Canto VI (Edited by Cesare Segre)	Canto VI (Trans. William Stewart Rose, with a slight modification in Octave 63)
son gioveni impudenti e vecchi stolti, chi nudi e chi di strane pelli involti.	With hoof of goat that other stamped the sand; While some seemed centaurs, quick in fight and rape; Naked, or mantled in outlandish skin. These doting sires, those striplings bold in sin.
62 Chi senza freno in s'un destrier galoppa, chi lento va con l'asino o col bue, altri salisce ad un centauro in groppa, struzzoli molti han sotto, aquile e grue; ponsi altri a bocca il corno, altri la coppa; chi femina è, chi maschio, e chi amendue; chi porta uncino e chi scala di corda, chi pal di ferro e chi una lima sorda.	62 This gallops on a horse without a bit; This backs the sluggish ass, or bullock slow; These mounted on the croup of centaur sit: Those perched on eagle, crane, or estridge, go. Some male, some female, some hermaphrodit', These drain the cup and those the bungle blow. One bore a corded ladder, one a hook, One a dull file, or bar of iron shook.
63 Di questi il capitano si vedea aver gonfiato il ventre, e 'l viso grasso; il qual su una testuggine sedea, che con gran tardità mutava il passo. Avea di qua e di là chi lo reggea, perché egli era ebro, e tenea il ciglio basso: altri la fronte gli asciugava e il mento, altri i panni scuotea per fargli vento.	63 The captain of this crew, which blocked the road, Appeared, with monstrous paunch and bloated face; Who a slow tortoise for a horse bestrode, That passing sluggishly with him did pace: Some here, some there, sustained the load, For he was drunk, and kept him in his place. Some wipe his brows and chin from sweat which ran, And others with their vests his visage fan.

The nature of the *intoppo* is that of a motley crew of magnificently horrific creatures, a veritable psychedelic reconfiguration of the discombobulated makeup of despondency, that threatens the progress that the good warrior Ruggiero is making towards safety. Like Ruggiero, our science, Galileo intimates, will only go so far as the pathway is open. The infinite variety of perturbations caused by the unpredictability of accidents that populate the universe of being cannot be subdued into human form. Although the good Ruggiero will inevitably triumph at the end, the force of the imagination may or may not succeed. Galileo and we must grant this much. But the memories of the good fortune accompanying the feats of Ruggiero dovetails with the astonishing power that the hermetic imagination bestows on human being-in-the-world. The horizon of visibility is expanded by the mood of reliance of the poetically trained imagination that is integral to the processes of knowledge. Being-in-the-world of hermeticism is willing to fly upwards towards the safety and silence of the God-maker. Galileo's imagination is suffused with the hermetic tonality that encourages the pursuit of knowing the world beyond the limitations that logic and experiment cannot overcome. He has always been portrayed as a champion of experience and geometrical demonstration, which is absolutely correct. But I would like to add to this picture the hermetic furor of the imagination, the hermetic *theologia gloariae*, that overcomes the *duro intoppo*, and without which the *nuova scienza* of uniformly accelerated motion cannot take off the ground. Galileo's new science, Ariosto's psychedelic poetry, and the hermetic imagination that amplifies the possibilities of understanding in despondent being-in-the-world are of one piece.

· 3 ·

LUNA

Received historiography has it that the publication of Galileo's *Sidereus nuncius* in 1610 shook the intellectual foundations of Europe, ushering in the age of the telescope.[1] The most spectacular announcement made by Galileo indicated that the moon, observed through a mysterious optical instrument, was mountainous and valleyed, in fact much more similar to the earth than hitherto believed. I suggest that the actual "shaking" was done not so much by the moon headlines and brilliant publishing operation mounted by Galileo, as by an intense, all but now forgotten debate that, in the aftermath of *Sidereus*, conjured early myths, unconscious visions of the cosmos, and philosophical undercurrents that had remained dormant in academic thought as practiced in early modern European universities. First, I will look at Galileo's colleague in Padua, Cesare Cremonini, the most revered Aristotelian natural philosopher in Europe. Cremonini exhumed (though not espoused) Dante's theological theory of the moon spots, as an indirect response to Galileo's *Sidereus*. Second, I will look at Giulio Cesare Lagalla, a professor at Rome, who portrayed Galileo as a Mercurius wielding a telescope-caduceus, as a second Trismegistus, and evoked an alchemical setting featuring a Galileo intent on

1 For the most up-to-date survey of the historiography and reception of *Sidereus*, see the excellent *Introduction* by Tiziana Bascelli and William Shea, in Galilei 2009.

the restoration of the philosophies of the ancient pre-Socratics. Finally, I will aim the spotlight at Fortunio Liceti, who challenged Galileo on the ashen light of the moon from the perspective of Aristotelian natural philosophy.

The historiographic canon presents a view that was epitomized already in the first biography of Galileo written by his pupil Vincenzo Viviani. However, Viviani also underlined the immense intellectual labor that was stirred by *Sidereus nuncius*. It is the institutional effects of this labor that have failed to capture the imagination of historians and philosophers. Viviani recounts that the "unexpected novelties published in *Sidereus nuncius* [...] gave the philosophers and astronomers of the day ample matter for discussion. At the beginning most of them felt repugnance for such novelties. Many were bold enough to rebel against them either with private communications or even in print since they thought them to be chimerical and preposterous, Galileo's false impressions, or possibly illusory appearances caused by the lenses. Shortly thereafter, though, all of them conceded to the opinion of the wiser and to sensory experience".[2] Viviani does not miss the disorientation, the reawakening of myths that precedes the instauration of a new intellectual order, that is, the shaping of a novel institution. His narrative condenses the tumultuous events, the creative spirit of resistance of the torch-bearers, of those who bring the enlightenment of the tradition, almost into a historic instant, after which 'wisdom' prevails, and sensory experience becomes the new law. It is in such historic instants, in their special temporal density, here marked by Viviani's laconic 'shortly thereafter', that the realization of institutions occurs. So it was for the institution of telescopic astronomy. The historiographic canon that we have inherited largely from nineteenth-century positivism has never emphasized the creative turbulence that preceded the laconic 'shortly thereafter'. When it did notice the stir it was from the perspective of the new order. This was inevitable as we cannot do away easily with the cultural orders that structure our existence. Yet we can work on them from within. We can embroider alternative motifs on the fabric of an established cultural order. We need not dismantle the canon, nobody can do that from within the horizon of their epoch, but we can make it less impenetrable by experimenting with it through a play of deviance, by asking illegitimate questions and refusing to answer them, and by letting the play present itself over and above the individual players. In what follows I will try to practice this imaginative historiography. The cabinet of curiosities that I will be restoring fascinates me. I was

2 Galilei 1890–1909, XIX, p. 610.

stunned when I found among the junk the dusty portrait of an improbable Galileo-Trismegistus.

Resurrecting Dante on the Moon

In 1613, Cesare Cremonini, Galileo's friend at Padua for almost two decades, published a *Disputation* on the heavens. The *Disputation* was a cosmological masterpiece in its own right. It dealt with the nature of the heavens, their motions, the separate motors of the celestial orbs, and contained two special tracts, in the form of apologias for Aristotle's opinions, one on the Milky Way and the other on the moon. Cremonini pointed out that his work on the moon was addressing one overarching issue, namely, the restoration of the correct reading of Aristotle's doctrines. These cannot be considered antiquated, says Cremonini, even in the aftermath of Galileo's fables. However, Cremonini's Aristotelian credo cannot be construed as a proclamation of agnosticism. It was rather a matter of the inquisitorial control of conscience that, after the Council of Trent, had succeeded in disciplining moral and intellectual life in early seventeenth-century Italy. On the other hand, Cremonini, the radical Averroist who denied the immortality of the soul, was chased by the Roman inquisition for years.[3] Had it not been for the academic-diplomatic 'immunity' accorded by Venice to the Padua professors, he would have ended his days burned at the stake, as Giordano Bruno did in 1600. The masquerade of a purely philological intent was not difficult to debunk, no inquisitor would have been fooled. At one point Cremonini cannot but shed the mask. He says: "Philosophical reason has to be grounded in the nature of the thing; once this has been scrutinized no difficulty or doubt will remain. Scholars have always pursued the nature of the thing. Whosoever misses the nature of the thing, he chases a ghost".[4] Cremonini reviews with great excitement the opinion he attributes to the Pythagoreans that was cautiously alluded to by Galileo, and according to which the moon is a second earth and possibly inhabited. He almost paraphrases Galileo, reporting that the Pythagoreans believe that the spots are asperities of mountains and cleavages of valleys similar to those we see on earth. But to avoid scandal he admonishes the reader that this opinion

---

3 Poppi 1992.

4 "Ratio philosophandi a rei natura petenda est; re inspecta nihil ardui relinquitur, nihil dubii. Sapientes enim ad rei naturam sunt loquuti; alii, quiqumque a re distant, ad genium qualemqumque". Cf. Cremonini 1613, p. 59 (the *Appendix* on the Milky Way and moon has its own separate pagination).

will be proposed by way of joking, as the Pythagoreans are notorious for telling fables rather than make philosophical arguments about fundamental things.[5] What they say about the ruggedness of the moon and its inhabitants must be fantasy land, unless they had an instrument with which they could see them, but nobody knows what this instrument would have been like, and at any rate it might have been deceptive.[6]

Thus, Cremonini convincingly caricatures Galileo (without ever mentioning him) as a fabulous Pythagorean turning his secret instrument to the moon but failing to see the lunar people. But was it really a caricature? Cremonini insists that fables are not philosophy, and he wastes no time discussing the telescope. Why is Galileo's ghost evoked as a story-teller who does not make proper philosophical arguments? The section of the *Disputation* discussing the face on the disk of the moon cast strange colors on the mountains and valleys and seas announced by *Sidereus nuncius* three years before. The question of the face refers to the shape of the dark spot that some pretend to discern with the naked eye in the disk of the full moon. But the natural philosophical issue that is at stake, the thing to be pursued by the philosopher, concerns the then widely received explanation of the dark face by differences in the density of the lunar body. The dark face was caused by the interaction of celestial light with a portion of the lunar body that, according to some was more dense, while according to others might have been more rare. Cremonini accepts this explanation as the correct one, consistent with Aristotle's doctrines, by arguing that the only way in which bodies interact with light is either by translucence, we might say by transparency, when they are less dense, or by effusion, here we might anachronistically paraphrase by radiation, when they are more dense. The body of the moon, situated within the celestial orb closest to the earth, is more impure and thus apt to shine by radiation. Hence the only remaining doubt is whether the dark face is the denser or rarer portion of the lunar body. Cremonini believes, logically, that the dark face is the rare portion of the moon.[7] Surprisingly, though he is not in agreement with Galileo's Pythagorean speculation that the brighter portion of the moon is made of earth while the darker part is made of water[8], Cremonini's conclusion harmonizes with the principle that the dense is apt to shine by radiation, since the earth is more dense, while water is more rare. He makes

5 Cremonini 1613, p. 50.
6 Cremonini 1613, p. 55.
7 Cremonini 1613, p. 64.
8 Galilei 1890–1909, III.1, p. 65.

sure to get it right on the nature of the thing, on first principles, for there lies the philosophical genius. The seventeenth century was the century of masks. Everybody wears their preferred mask to speak their voices on the stage of the world. Cremonini likes posturing as Aristotle, Galileo plays Pythagoras.[9] The former pokes fun at the latter's fables, and the latter derides the madness of the former who pretends not to know that the mysterious instrument with which the Pythagoreans observed the moon was recently recovered.

Galileo hinted at Pythagorean fables but Dante, three centuries before Galileo, had put forward a theory of the dark face of the moon that could not be swept aside as mere fantasy. Cremonini is drawn to Dante's theory since there, for him, shines the genius of the poet (*…quia tamen praestat videre poetae ingenium*).[10] The question that strikes Cremonini is whether poetic *ingenium* has access to truth beyond the discourse of philosophy, that is, beyond the logos of explanatory reasoning. Pursuing this question is all the more imperative, says Cremonini, as the *ingenium* of the poet, poetic mind, departs from received reasoning (*…quo magis recedemus a communibus explicationibus*).[11] Dante let loose his speculative genius in Paradise, the third *Cantica* of the *Commedia*. Accompanied by a celestial Beatrix, the spirit of the woman of his youthful love, Dante begins the ascent to the *Empireum*, the heaven that is also the abode of God. He traverses all the spheres of the medieval cosmos, where the spirits of the blessed orderly appear to him in a succession. While transiting on the sphere of the moon, Dante asks Beatrix an explanation for the appearance from earth of the lunar dark face. His tentative opinion is that it must be due to some difference in the density of the celestial matter, the then

9 On the question of the extraterrestrial inhabitants of the cosmos, Galileo publicly claimed (*Letters on sunspots*) that the opinion of those who think that on Saturn, Jupiter, or the moon, there might be animals like those living on earth is surely false and to be condemned (as theologically unacceptable, he seems to imply). However, he was agnostic as to the possibility that both animals and plants whose nature is far removed from our imagination might be living on other planets (Galilei 1890–1909, V, pp. 220–221). Once the caution of the public pronouncement is taken into account, I believe there remains little doubt about what Galileo really thought. It seems to be no less theologically dubious to believe in alien forms of life than in the life of humans or other known animals in other parts of the cosmos. Yet the book was published with an enthusiastic imprimatur in Rome advertising the excellent natural-philosophical doctrines put forward by the author with commendable clarity of style. "…immo raram doctrinam, novas ac mirabiles observationes hucusque incognitas inauditasque, facili ac perpolito stilo explicatas, continere invenerim" (Galilei 1890–1909, V, p. 74).

10 Cremonini 1613, p. 78.

11 Cremonini 1613, p. 78.

received explanatory framework. But Beatrix corrects Dante's misguided opinion and launches into a vastly more ambitious attempt to relate the presence of the dark face of the moon to the mysterious influence of the divine mind on creation.[12] It is a theory that combines theology and physics with almost a mechanical flavor. It correlates the percolation downwards of divine influence with the different inner structure of the bodies of the heavens that receive the power emanating from the first intelligence. The Canto culminates with a fascinating metaphor hinting at a metaphysics of light. As human happiness shines through the opening of the pupil, so the joyful nature of the divine penetrating body makes it shine. The divine virtue mixing with body becomes light itself, like the transmutation of matter into spiritual quintessence of an alchemical opus. It is the mysterious efflorescence emanating from the joyful human face that reveals the presence of happiness in a human being. So with the divine being. It is light in body, matter made luminous, that makes the mysterious presence of the divine perceptible. One has almost the impression that Dante is willing to make God an object of sensory experience. The dark face of the moon is not dark. It is the lunar mode of reception of the divine power within the moon's body that makes it appear to us as colored. Cremonini has dutifully defended the naturalistic explanation of the lunar dark face that fits the Aristotelian framework of scientific knowledge. Yet he is almost struck dumb by the audacity of Dante's genius. Cremonini has no ammunition to rebut Beatrix's holistic vision. He concludes his disputation with a wishy-washy section entitled "Considerations", criticizing Dante's theory on matters of detail. It is instructive to note that Cremonini's rediscovery of Dante's lunar theory in the wake of Galileo's *Sidereus* was unforeseen, as Cremonini himself admitted that he had never read Dante's poem until then. The *Sidereus nuncius* really worked wonders, as literally as the book's title implied that it should. The cosmological announcement made the reawakening of ancient mythologies and theological visions part of the choral event that sounded the death-knell to the dictatorship of academic science. It was also the prelude to the realization of a novel institution, telescopic astronomy, which was not based on first principles of science but, grotesquely, on possibly deceiving instruments, and myths and fables recounted on the stage of the world by characters wearing masks.

---

12 Alighieri 1987, pp. 786–791.

Hermes and the Telescope

In 1612, two years after hitting headlines with the first scientific bestseller in modern science history (Galileo's *Sidereus*), the Venetian publisher Tommaso Baglioni printed another book on the moon. The author was Giulio Cesare Lagalla. The title summarized Lagalla's manoeuvre. "Physical disputation on the phenomena on the disk of the moon that have recently been stirred up by Galileo Galilei by means of a new telescope…to which a second disputation is added on light and luminosity".[13] The phrasing of the title alludes to the conjuring of phenomena out of the telescope. Lagalla avoids saying that Galileo discovered new phenomena with the telescope. He stresses that phenomena were stirred up, conjured out of the telescope by Galileo as if by magic. The setting of the telescope in a magical context must have been carefully thought out by Lagalla. He had been introduced to Galileo in 1611 when the latter showcased his instrument in Rome. Galileo and Lagalla became close friends. They made observations with the telescope together, a circumstance which gave Galileo the opportunity to raise further questions concerning light. He showed Lagalla and the Roman intelligentsia a luminous stone, a mineral recently discovered in the Bologna region. The stone was a preparation of barium sulphate that became luminescent in darkness after having been exposed to sunlight.[14] Lagalla understood the magical-alchemical context of the phenomenon that Galileo had demonstrated to him. He also understood that Galileo aspired to present himself as the bearer of the light that would illuminate the darkness of natural philosophy. Since its casual discovery the Bologna stone had been regarded as an alchemical mystery. The stone in its raw state was not luminescent. Some preparation was needed that was generally described as calcination. But there were many recipes, and contemporary accounts suggest that the preparation process became an art in its own right.[15] Who calcined Galileo's stones? Had he 'dirtied his hands' with alchemy? Was he looking for the stone? Perhaps Galileo learned how to master the alchemical transmutation of the raw stone, the secret opus that would make the stone luminescent, and/or was able to obtain prepared specimens.[16] The association

13 Lagalla 1612.

14 Cf. Roda 2011, for the early history of this discovery.

15 Cf. Cellio 1680, for an account of the recipes.

16 There is no question that the Bologna stone was immediately seized upon as an alchemical marvel. A Bologna poor shoemaker with an interest in alchemy, Vincenzo Casciarolo, discovered the stone in the hills surrounding the city. He cooked the specimen and after

of magic, alchemy, and telescopic astronomy in Galileo's early Roman demonstrations is reflected in the portrait of Galileo painted by Lagalla in the introductory section of the 1612 *Disputation*.

Here is the beginning of Lagalla's mythography. "Learned antiquity portrayed Mercury wielding the caduceus as a messenger of divine things and as the savior of the souls of the dead from the lower regions. Our ingenious age witnesses and admires Galileo, novel interpreter of Jupiter, opening up the heavens with a caduceus-telescope, reawakening the ghosts [manes] of the ancient philosophers. Indeed, as both the name and function of this sidereal messenger are the peers of those of the ancient Mercury, so Galileo's telescope does not differ from Mercury's caduceus. Setting aside the shape of a tube in the ancient representations of the caduceus, perhaps hinting at the telescope, I marvel at the snakes coiling around it and staring at each other on its top, unequivocal symbol of the perspicuous and vigilant eye that is attributed to them, it being strangely reported that some animal may paradoxically be killed by their gaze, a fact that though repeated by many is believed by very few, and so this telescope reinforces the acuity of vision so much that it would hardly be believable or even fabulous".[17] Whose demons are those reawakened by

---

exposing it to the sun he found out that the stone would give off light in darkness. He thought the stone might hold the secret to transmute matter into gold and promptly passed it on to a professional alchemist (Cellio 1680, pp. 33–35). Christian Mentzel, a student of Fortunio Liceti in Bologna and subsequently at Padua (see below, for Liceti's involvement with Galileo), compared and contrasted the luminescent stone with the *hermetic phosphor* described by Christian Adolph Balduin (Balduin 1675, Keller 2014). See Mentzel 1675, in the *Dedicatio*, not paginated, who tells us that Liceti was his 'promoter' in the medical faculty, at Padua I suppose, where Liceti had moved from Bologna, and that he had lived in Bologna busying himself with the stone, collecting samples, and perfecting the calcination process. Although he discussed the Bologna stone from the viewpoint of Aristotelian natural philosophy, Mentzel acknowledged that in the correctly prepared stone an analogical fire (...*analogicum aliquem ignem*) was instilled that fit the analogy between the macrocosm and microcosm (Mentzel 1675, *Proemium*, not paginated). The analogical fire implies the mixture of celestial substances or light with terrestrial matter.

17 "Mercurium caducaeum gestantem colestia nuntiare, et mortuorum animas ab Inferis revocare sapiens finxit antiquitas: Gallilaeum vero novum Iovis interpretem telescopio caducaeo instructum Sydera aperire, et veterum philosophorum manes ad superos evocare solers nostra aetas videt, et admiratur, neque vero, ut par ei nomen, atque officium cum Mercurio est Syderei Nuncij, a prisci illius Mercurij Caducaeo eius telescopium differt; omitto Iubae figuram, quam in aniquissimis Mercurij signis conspicimus, huius fortasse instrumenti formam praesignantem, sed potissimum contemplor serpentes caducaeum amplexos: mutuo se in summitate aspicientes, certissimum vigilantis,atque perspicacis oculi symbolum, cuius tanta in his animalibus vis creditur, ut inopinabili eventu aliquod

Galileo-Mercurius? Lagalla mentions the mythical Orpheus, the pre-Socratic Thales and Anaxagoras, the Pythagorean Philolaus, the atomist Democritus, Heraclides Ponticus, and Plutarch's Lamprias (a reference to the main character in the latter's well-known dialogue on the face appearing in the moon and a supporter of the Plutarchean-Platonic worldview).[18] Let us set aside here the doctrinal implications of this genealogy for the lunar novelties and focus on the imagery of Galileo-Mercurius. There is no doubt that Lagalla hit upon a seductive metaphor. Its implicit amplifications evokes a series of mythological images in the reader's mind. The telescope-caduceus points to the significance of the Latin name of the rod wielded by Mercurius. Its meaning was expounded by Vincenzo Cartari in his widely circulated mythology textbook.[19] He argued that the Latins named it caduceus because at its sight all discords disappeared, and thus it became a symbol of peace.[20] Was the newly invented telescope, Lagalla rhetorically asked, going to make all philosophical discords cease by virtue of the lenses, like the snakes's eyes staring at each other? Perhaps Galileo had initially deluded himself that this should have been the case, for we know that, long before his friend and former colleague Cremonini published the *Disputatio*, he complained that he despaired of winning over the stubborn Aristotelians.[21] Yet the metaphor of the caduceus was loaded with ambiguous meanings. The rod acquired the coiling serpents for the first time, says Cartari, when Mercurius threw it among two snakes fighting with each other. They ended up coiling around the rod with their heads on top fixed in a state of suspended animation.[22] The fight was over but had there been a winner? Likewise philosophical disputes may be terminated with peace treaties and yet remain formally undecided as to their truth value.

The most powerful image evoked by the metaphor is the figure of Mercurius as Hermes Trismegistus. Galileo is thus saluted by Lagalla as a hermetic sage in the twilight of that tradition. Mercurio Trimegisto, the name of the ancient Egyptian high priest in Benci's rendition of Ficino's *Pymander*, possessed three essences, as I have already mentioned. He had received within himself the image of God the father (referred to as *the abyss of piety and mercy*),

---

ex ipsis, conspectum animal interimere posse perhibeatur, quod veluti certum a multis traditur, et a nullis, aut vix paucis, creditur, ita huius telescopij vis tantam videndi aciem profert, ut supra hominum credulitatem narratum fabulose audiretur" (Lagalla 1612, p. 1).

18 Lagalla 1612, p. 1.

19 Cartari 1581. The book went through countless editions.

20 Cartari 1581, p. 263.

21 Galilei 1890–1909, X, pp. 423, 484.

22 Cartari 1581, p. 261.

the abyss of the Son's learning, and the gift of prophetic spirit.[23] Now we can see that this association between Mercurio and Galileo's *Nuncius* was made explicit by Lagalla. It was a possibility of understanding the ongoing transformation in their philosophical outlook that was open both to Galileo and his friend. Being-in-the-world finds itself immersed in possibilities of understanding that exist before the separation of mind and world, and which are prior to the emergence of a distinction between theory and facts of observation. The emblematic association of caduceus and telescope belonged in Galileo's life-world. In the introduction to the text, Ficino claims that Trismegistus, among the greatest Egyptian philosophers, had raised himself from mathematics and natural philosophy to the contemplation of divine things. But to understand the divine mind the human soul must first become infused with divine light. The soul cannot achieve this state if she does not totally absorbs the divine mind by turning towards it as the moon turns towards the sun. This turning of the human mind can only be achieved by forsaking both the senses and fantasy. The moving away from perception and imagery prepares the way for the divine mind to flow into the human soul.[24] There is implicit in this extended simile the idea of the moon being illuminated by the sun. It is Galileo's view that the moon does not shine of its own light but being rather an opaque body it reflects the light of the sun. Yet there is an illuminating paradox here. To contemplate the divine mind, the very order of creation, to reveal the mystery of the lunar dark face, the human soul must forsake not only the senses and hence the visual experience of the telescope as well, but the whole *phantasmic* universe of the Renaissance.[25] Thus the gates of the realm of wisdom will be flung open on the verge of the abyss of the Son's learning. Knowledge is the frozen gaping at the abyss that is symbolized in the suspended animation of the serpents. The epistemic conundrum raised by Lagalla's potent metaphor is that enlightenment in the human soul cannot be accomplished without surrendering our yearning for the phantasmic and the experiential. This is the condition, even for Galileo, the second Trismegistus, to receive the abyss of the Son's learning.

---

23 Trimegisto 1549, in the section *Testimonianza di Raziel sopra Mercurio Trimegisto* (not paginated).

24 Trimegisto 1549, in the section *Argumento di Marsilio Ficini Fiorentino*...(not paginated).

25 Cf. Couliano 1987, who speaks of the phantasmic mind characterizing the Renaissance.

Demodocus and the Ashen Light of the Moon

A native of Rapallo, near Genova, Fortunio Liceti was among the most productive, ingenious, and versatile natural philosophers of the early seventeenth century. In 1653, he published a list of his works, some in print and some forthcoming. The publications include nine works on philosophy, ten on medicine, eighteen on mathematics and physics, and fifteen of a philological character. In total, he lists fifty-three books. Another section includes the forthcoming books.[26] Yet no serious book-length study of this great master has ever been produced over the last four centuries.[27] Under the spell of nineteenth-century positivist incantations, in a book still recounting the soothing fable of Galileo, the father of modem science, Liceti was dismissingly called "a verbose philosopher".[28] More perceptively, Don Cameron Allen noted the immense erudition of Liceti's "symbolic analogues and interpretations" insisting that they are based on "enormous learning".[29] Indeed, his learned originality transpires, for instance, from the fantastic 'ethico-physical' dialogue inspired by the Homeric episode of Odysseus and Circe concerning the question of the scope and limits of naturally occurring transformations of bodies.[30] Liceti's lifelong friendship with Galileo, and above all the late controversy with the Florentine astronomer over the ashen light of the moon, may have unjustly obscured his brilliant contributions to the philosophical culture of the time.

In 1640 Liceti put forward the bold thesis that the ashen light of the moon is a lunar atmospheric phenomenon due to a luminescence effect similar to the Bologna luminous stone. In *Sidereus nuncius* Galileo had argued that it was lunar 'earthshine' caused by solar light reflected by the full earth on the moon. When Galileo wrote his response to Liceti, with the help of an assistant, he was blind and suffering from memory impairments, hence he made a plea for compassion that sounded a little pathetic.[31] However, Liceti paid reverence to Galileo's condition by comparing him to Demodocus, the poet at the court of Alcinous who entertained Odysseus with his songs.[32] Like Demodocus, says Liceti, Galileo had received from the muse of wisdom the

26 Cf. Liceti 1653, catalog at the end of the volume.
27 Ongaro 2005 offers a short biographic profile and an exhaustive bibliography.
28 Piccolino & Wade 2014, p. 56.
29 Allen 1970, p. 275.
30 Liceti 1636.
31 Galilei 1890–1909, VIII, pp. 467 ff.
32 Liceti 1642, p. 55.

good of the mind and the evil of corporal illness, namely, blindness. But Liceti humorously noted that for all his postulations, Galileo still appeared to have retained some pretty good memory for learned quotations and for observations pertinent to the subject matter. Liceti concluded that judging from the cogency and articulation of his essay, Galileo was not at all decrepit, as he wanted the readers to believe, but rather in that ripe old age to which Lucretius had ascribed the greatest wisdom and the strongest moral force.[33] As for the memory impairments, a dollop of Aristotelian cognitive science sufficed to question Galileo's plea for compassion. Liceti pointed out that memory is a habit of the faculty of fantasy, and that Galileo had certainly not lost his knowledge, or *scientia* which is, according to Liceti, a habit of the intellect. Besides, Liceti suggested, those who are better endowed in terms of their judgment tend to have worse memory performance and vice versa. Galileo's mind would not have been subject to the natural process of decay that perishing bodies undergo. Liceti had proposed the same argument in his dialogue on the nature of transformation. There he had the sorceress Circe explain to Odysseus that she could transform humans also into plants by natural magic (having them retain the faculty of hearing yet depriving them of their speaking organs, hence of the faculty of linguistic communication), but that she had no power over their minds and passions, for mind was only subject to divine power.[34] To an amazed Odysseus Circe clarified that her plant-humans still communicated by movement though not through ordinary language, thus suggesting that human language need not be the exclusive medium for conveying thought and passions.

Liceti's book on the Bologna stone, where he first published his speculation on the ashen light of the moon, was entitled 'Litheosphorus', a Greek-sounding neologism for a work that summed up his natural philosophical learning in the wake of his perennial 'teacher', not a person, but to a large extent the late medieval tradition of the so-called Latin corpus of Aristotle.[35] In chapter fifty he drew a startling analogy between the body of the moon imbibing and retaining solar light and the Bologna stone. Two years later, he addressed Galileo's opposition by urging him to consider the *Litheosphorus* as no more than a little work in the vein of the Platonic or Plutarchean *Symposia.* His goal was to offer savory food for thought, not excessively rich in condiments, so

33 Liceti 1642, p. 56.

34 Liceti 1636, p. 10.

35 Liceti 1640, pp. 247–248.

as to satisfy the hunger of the reader by following correct Galenic dietetics.[36] The metaphor of food is pregnant with meaning. Evoking images of hunger and well-balanced eating Liceti emphasized that natural philosophy is not reducible to dialectics. Rather, it is the healthy physiological processes of digestion and assimilation of ideas that concur in the formation of good science. Liceti was right if one judges by the two works he produced in relation to this topic. The *Litheosphorus* is brimming with ingenious material that transcends disciplinary barriers. It can be read as an extended inquiry into the nature of luminescence as a first principle of nature. The *De lunae subobscura luce*, his formal response to Galileo, is overall less brilliant, betraying the fatigue of a dialectical rebuttal of Galileo's tract point by point, and appearing to have been dictated more by professional egotism than by sincere interest in further pursuing the topic.

But what did Galieo find questionable in the Bologna stone metaphor for the moon light? Liceti claims that since moonshine appears so bright at full moon, we should expect that if the ashen light were due to the earth's reflection of solar light when the moon is at inferior conjunction, i.e., at new moon, it should be much brighter than the pale light which is observed, which is less bright than moonshine. This, he argues, is both because the terrestrial globe is greater than the moon, thus reflecting more light, and because the moon at superior conjunction is further from the sun than the earth is when the moon's phase is new, i.e., at inferior conjunction, and therefore the moon's reflected light must be weaker than the earth's reflected light.[37] Galileo answers that this expectation is exactly right, and that the ashen light of the moon at inferior conjunction is indeed much brighter than moonshine at full moon. Further, we would observe this difference as expected, were it not for the fact that our senses subtly deceive us in this particular case. But, Galileo asks, how is this weaker ashen light actually 'seen'? First, Galileo notes that the same bright object illuminating an opaque object will light it more or less according to the distance separating the two of them. Thus the same object will appear less bright when placed at a greater distance. Consequently in comparing the moonshine and the lunar candor (as Galileo refers to the ashen light) we should keep in mind that the earth lit by the moon is no further from our eyes than three or four braccia, whereas the ashen light comes from the moon at inferior conjunction, which is separated from our eyes by more than three hundred million braccia.

---

36 Liceti 1642, p. 60.

37 Galilei 1890–1909, VIII, p. 484.

Obviously, then, we cannot marvel at the fact that even though moonshine and lunar candor were equally luminous they would appear different to us. Before drawing any conclusions, Galileo says, Signor Liceti should have first noted down how much light is due to moonshine here on earth and then gone to the moon to check how much light due to earthshine is actually there. To obviate the difficulty of a trip to the moon Galileo suggests that a comparison should be made with a third type of light. The light at sunset is, in Galileo's view, apt to make that comparison. After sunset we can still distinguish minute things which could not be discerned with the light of moonshine. Furthermore, an object in the distance remains visible in the twilight after sunset for a long time, but will no longer be visible with the light of moonshine. Thus, Galileo continues, moonshine must be less bright than twilight. On the other hand, in order to see the lunar candor we need not wait the extinction of twilight, and the setting of complete darkness. The lunar candor is in fact perfectly visible against the backdrop of twilight. Therefore, Galileo concludes, twilight is less bright than the lunar candor, and moonshine must in turn be less bright than the lunar candor.[38] The blind Galileo has not lost the capacity for imagining luminosity differences. 'Brighter' and 'less bright' are still vivid in Galileo's imagination. He has recollections of past experiences, of course. But Galileo also rewrites his memory. He projects himself as an experimenter in a mental scenario populated by silhouettes, such as the unspecified 'minute things' still discernible in the twilight. The narration of imagined experience is a tool for understanding the physical world. At one point Liceti argues that some of the cavities observed with the telescope by Galileo on the surface of the moon should function as mirrors, thus reflecting and reverberating solar light in the moon's ether. This would make the lunar candor in the central part of the moon brighter than at the periphery, an effect which, according to Liceti, is not observed.[39]

Galileo replies as follows. First of all, we know that concave mirrors reflect light but do not illuminate the surrounding air. They light opaque bodies which are hit by the reflected rays. They do so with much more force than the direct light of the sun, as burning mirrors show. Indeed, concave mirrors concentrate light at the focal point, the cusp of the projection cone. But to see the reverberation of light in the surrounding air we need to have the reflected rays meet an opaque matter. Imagine some matter suitable to produce dense smoke, such as incense, above burning charcoals placed right below the cone

38 Galilei 1890–1909, VIII, pp. 513–514.

39 Galilei 1890–1909, VIII, p. 484.

of light emitted by the mirror. Then you will see the cone lit up because of the illumination of the smoke. Now, Galileo goes on, even though those cavities hypothesized by Liceti reflected the solar rays as mirrors, we could not see them anyway, because all we would see is the back of the opaque matter which would not be lit up by the rays. Furthermore, even if the moon were an entire concave mirror, a small portion of the concave lunar orb, i.e., a portion of a spherical surface whose radius were equal to the distance between the centre of the earth and the lunar sphere, the only reflected light coming down to us would be concentrated in the vertex of the projection cone. Thus this concentrated light would only occupy a minute particle of the terrestrial surface. Here it would eventually become visible, but only just so, given the very modest curvature of such a cosmic mirror. So, even if Liceti decided to look for this tiny spot, to corroborate his argument, he would be well advised not waste his time in so vain a pursuit.[40] Galileo has given Liceti a short lecture on concave mirrors, only, ironically, to send him roaming the surface of the earth, in search of the speck of faint light that the moon-like-a-mirror would reflect towards us. For Galileo, who, as we know, was a great admirer of Ariosto's epic poem, *Orlando furioso*, the image of a philosopher roaming the earth in search of light might really have mirrored the image of the Paladin Astolfo wandering the moon's valleys, where, according to Ariosto, all that is lost by humans ends up. Astolfo went to the moon in search of the wisdom of the poem's hero, Orlando, who had gone mad because of love. At one point Ariosto lists the many ways in which humans lose their wisdom here on earth. He pairs astrologers with philosophers in their hopeless enterprises. Notwithstanding the augural meaning of his first name, Fortunio (lucky), Liceti might pointlessly spend the rest of his life chasing the impossible spot of light.[41]

Both Liceti and Galileo were humanists through and through whose inflated egos, in the heat of the controversy, lost sight of the autonomous energy of metaphors. They unconsciously projected humanistic archetypes on the mystery of the ashen light by deploying metaphors to illuminate first principles. For, in the tradition of Italian humanism, it was metaphor that gave philosophical discourse the depth of meaning that could found rational inquiry about first principles.[42] The latter are indemonstrable yet indispensable for deductive demonstrations to get off the ground. In highlighting the philosophical import of Italian humanism, Ernesto Grassi pointed to

40 Galilei 1890–1909, VIII, p. 521.

41 Cf. *Orlando furioso*, octave 85, canto 34 (Ariosto 1999).

42 Grassi 1983.

the fundamentally *pathic* dimension that informs the human experience of reality.[43] Recall that in Ficino's *Pymander* humanistic ideals and hermeticism coalesced, and Trismegistus was portrayed as the recipient of the abyss of the Son's learning. Likewise Grassi has spoken of the "impelling call of the abyss, by virtue of which reality comes to originary presence".[44] The abyss of the Son's learning can be sounded by metaphors that carry the mysterious flow of empathy through the Renaissance cosmos. Dialectical skirmishes aside, both the Aristotelian Liceti and Galileo-Demodocus, from the abyss of his terminal blindness, empathically shared an alchemical-hermetic metaphor of the moon as an opaque or quasi-opaque body, formed by a solid core that perhaps had a spongy nature and was surrounded by an ether. It was assumed as an indemonstrable first principle not resolvable by Galileo's telescope, that is, not by observation alone. The lunar corporeal nature could only be expressed in human language by the metaphor of a dark matter illuminated by the sun. It needed to be established first, in order for scientific, i.e., deductive, knowledge to proceed according to received methodological rules.[45] Fecundation of the lunar dark matter by sunlight mysteriously produced tenuous luminous effects, the ashen light, possibly by mirror-like reflection of sunrays by the earth, as Galileo would have it, or by luminescence such as in the calcined Bologna stone, as Liceti would prefer. Such conjugation of sun and moon, in its myriad manifestations, was a recurrent theme in the alchemical literature widely circulating in the Italian Renaissance and especially at the Medici court in Florence.[46] In the *Studium consilii coniugii de massa solis et lunae*, for example, the sun is said to be shining above all things, while the moon shines of an alien light since she borrows almost all of her splendor from the sun.[47] But the *coniunctio* of the two substances meant for the alchemists reciprocal annihilation so that a truly new nature, an ontological 'otherness', could be born. Cryptic and obscure as they appear to us uninitiated, the fundamental principles of alchemy were deliberately concealed by the practitioners under the veil of metaphor and simile. Alfredo Perifano has shown that this was the common and accepted practice of the time.[48] Liceti knew that the inquiry

43 Grassi 1994, p. 55.

44 Grassi 1994, p. 55.

45 "…me noluisse discedere a methodo…ne progrederemur ex ignotis" (Liceti 1642, pp. 60–61).

46 Cf. the learned study by Alfredo Perifano (1997) on the widespread alchemical interests throughout Renaissance Italy.

47 Ars chemica…1566, p. 57.

48 Perifano 1997, p. 31.

about principles is not reducible to syllogistic deduction. It has to be driven by metaphor and analogical discourse.

The *Litheosphorus*'s alchemical hypothesis of the luminescent origin of the ashen light belongs in the same ideational framework that had been set in motion by Galileo's telescope. Notably, Vincenzo Casciarolo, the discoverer of the Bologna stone in the hills surrounding the city, was a dilettante alchemist. The association of the humble stone with the avant-garde, jagged moon seen through Galileo's telescopes catalyzed an intense societal preoccupation, learned and mundane, with the mysterious natures of light and matter. It was instrumental in activating latent archetypes of unity and transformation in Renaissance culture that shaped the realization of telescopic astronomy, a cultural 'otherness' whose identity remained concealed from the consciousness of the motley crew who labored at its founding.

I wish to conclude by reasserting the 'sacrificial' labor of resistance done by the exponents of tradition in view of our reappraising institutional aspects of historical change and thus addressing the distortions brought about by reductionist interpretive frameworks. The realization of a human institution is a principle of cultural becoming whose effects persist long after its inception, though its dynamics will soon be forgotten in the histories written by the survivors. The cultural labor that leads to the individuation of the structures and forces, both social and cognitive, that bring to life cultural objects is not done by the trail-blazers alone or by lobotomized social organisms. Individuals and social mechanisms as isolated entities are chimerical abstractions. The deregulatory activities of the innovators must be reinterpreted and questioned in the light of cultural traditions. This reconstructive brickwork with the mortar of myths and symbols and metaphors brings about the recognition of the archetypes buried in a culture. The depths are thus sounded of the unconscious formations of the social mind that shape the values of an entire civilization. The creativity is fostered that will help us rethink positivist images of science that fail to account for its creative mythological and poetic background.

· 4 ·

SOL

David Wootton's discussion of Galileo's religious beliefs, or unbelief, as he argues, has drawn attention to important documents that help us reconstruct Galileo's hermetic being-in-the-world.[1] Particularly relevant to my project is

1 Wootton 2010, pp. 244–247. In 1639, Galileo appeared to be desperately ill. When his life-long friend Benedetto Castelli received news of his bad health in Rome, he wrote him a letter, which seems to imply that Galileo had made statements in conversations, which Castelli thought to be divinely inspired (…*non possono venire se non da altissima banda*, Galilei 1890–1909, XVIII, p. 57), and from which Wootton deduces that Galileo had "given his soul to Christ" at the last hour (Wootton 2010, p. 247), as a phrase by Castelli clearly refers to the parable of the laborers in the vineyards. The letter is very cautious in detailing facts that might be compromising. Next, Castelli compares Galileo's resolution with the devotion of one Mother Sister Elisabetta, much talked about in Rome, who dedicated herself piously and strenuously to *imitatio Christi*. Castelli says that he is pleased to hear that Galileo is following the example of Sister Elisabetta. However, the crucial information about Galileo's statements that Castelli relays in the letter was not first hand. Galileo wrote nothing about this to his pupil in Rome directly. It had been passed on to Castelli verbally or in writing (it is not clear) by two intermediaries. Wootton's overall argument seems to me convincing. It would be interesting to speculate whether Galileo's alleged conversion to Christ at the last hour might tell us something about his previous convictions and/or doubts in relation to the divinity of Christ, a question that would fit the theological debates of Italian hereticality.

the testimony by Balthasar de Monconys, a Frenchmen who visited Florence in 1646. He reports the gist of a conversation with Galileo's disciple, Vincenzo Viviani. "Il me dit son opinion [Viviani's or Galileo's ?] du soleil qu'il croyoit une estoille fixe, la necessité de toutes choses, la nullité du mal, la participation de l'ame universelle, la conservation de toutes choses".[2] I agree with Wootton that although the grammatical ambiguity of the pronoun 'son' does not allow us to decide whether the opinions were Galileo's or Viviani's, contextual evidence, based on Viviani's close relationship with Galileo and on Viviani's personal religious convictions, suggests that it is quite possible that Monconys's passage relates Galileo's opinions, or that the opinions were shared by both Galileo and Viviani.[3] Monconys's journal shows that he was keen on meeting the Galilean younger pupils, and on collecting scientific information, books, especially Galileo's, and practical know-how about lenses suitable for telescopes and microscopes. The journal is also important for another reason. It makes the reader feel like she is touring Florence and Pisa at a time when the real action was over, and what she is left with is melancholic sightseeing, paying homage the glories of the Medici of a better time. The picture that emerges from Monconys's impressionistic diary is one of decadence, and Galileo memorabilia are offered on display for curious tourists. Galileo's body is shelved in a secret place waiting for the Medici to pluck up the political courage to challenge Rome.

The Sun

The reference to the starry nature of the sun and to the world-soul leads us to Galileo's biblical exegesis once again. Galileo's life-world implies a commitment to biblical exegesis, and his can be gauged by glancing at how he answered Cardinal Bellarmine's objection regarding Psalm 18. The verses in question are as follows: "High above, he pitched a tent for the sun, who comes

2 "He told me his opinion about the sun, which he thought to be a fixed star, the necessity of all things, the inexistence of evil, the participation of the universal soul, the conservation of all things" (De Monconys 1665, p. 130). This chapter in Monconys's travel journal is a lovely account of the social and intellectual milieu among Galileo's friends and pupils in Florence after the death of the master.

3 I also agree with Wootton that the marginal note in Monconys 1665, p. 130, referring the opinions to Viviani, must be considered indecisive in that Monconys's book was edited and published by his son posthumously, and therefore the marginal note might not have been in the manuscript diary.

out of his pavilion like a bridegroom, exulting like a hero to run his race. He has his rising on the edge of the heaven, the end of his course is its furthest edge, and nothing can escape his heat".[4] Monsignor Piero Dini had told Galileo that Cardinal Bellarmine would gladly see what he had to say on these verses, traditionally believed to attribute motion to the sun. The result of Galileo's exercise was expounded in a letter to Dini. First of all, Galileo expanded on the existence in nature of a calorific spirit distinct from light.

> [T]here is in nature a very spirited, tenuous, and fast substance which spreads throughout the universe, penetrates everything without difficulty, and warms up, gives life, and renders fertile all living creatures. It also seems to me that the senses themselves show us the body of the sun to be by far the principal receptacle for this spirit, and that from there an immense amount of light spreads throughout the universe and, together with such a calorific and penetrating spirit, gives life and fertility to all vegetable bodies. One may reasonably believe that this is something different from light since it penetrates and spreads through all corporeal substances...It seems to me that from the Holy Writ we can acquire evident certainty that the solar body is, as I have said, a receptacle and, so to speak, a reservoir of this spirit and this light, which it receives from elsewhere...For we read there, before the creation of the sun, about the spirit with its calorific and fertile power 'skimming the waters or lying over the waters' for future generations; similarly, we have the creation of light on the first day, whereas the solar body is created on the fourth day...Although I could produce many quotations from philosophers and serious writers in favour of the wonderful force and energy of this solar spirit and light, I want to limit myself to a single passage from the book of The Divine Names by the Blessed Dionysius the Areopagite. It is this: 'Light also gathers and attracts to itself all things that are seen, that move, that are illuminated, that are heated, and in a word that are surrounded by its splendour...'. [...] This philosophical view is perhaps one of the principal doors by which one may enter the contemplation of nature.[5]

After this preparation, Galileo goes on to put forward his interpretation of the entire biblical passage cited by Cardinal Bellarmine. In doing so, being-in-the-world of hermeticism projects further possibilities of understanding by exploring the meanings of the truth of the Psalm and the truth of recently discovered phenomena concerning the sun. One might be tempted to separate the two truths. There is the natural-philosophical truth concerning the sun.

4 Psalm 19:4. A number of manuscript writings in defense of Copernicanism, which Galileo widely circulated at the time, have been published under the heading *Scritture in difesa del sistema copernicano* [Writings in defense of the Copernican system]. See the originals in Galilei 1890–1909, V, pp. 261–370, and the English translation of the entire dossier in Finocchiaro 1989, pp. 47–118.

5 Finocchiaro 1989, pp. 63–65.

Then there is the religious truth stated in the Bible. But this separation would be totally anachronistic. It would be a reification of the positivist ideology of the separation of science and religion. In Galileo's life-world this separation is impossible.

Furthermore, it is not only the case that it would be incomprehensible to him and his contemporaries. Indeed, projection of understanding in the life-world seems to be a primordial mode of being which is prior to the phenomenon of separation of the modalities of understanding. The separation looks more like a cultural conditioning that is historically variable, which explains why Western culture in the twentieth-first century tends to demarcate natural philosophy from religion, in sharp contrast to the early modern period. It is the standard indoctrination that is recycled in educational systems, sponsored by and subservient to ruthless economic and financial powers, the corporate state governing the industrialized world. It peddles to students two cartoons, the fiction of the immaculate conception of science, born of human emancipation from obscurantism, and the fiction of religion and theology and myth as irrational and anti-scientific systems of beliefs.[6] The fact is that throughout early modern European history…*scientiam naturaliter religiosam esse nemo non videt*.

> So I am inclined to believe that the passage in the Psalm could have this meaning, namely, that God 'pitched a tent for the sun' (*Deus in Sole posuit tabernaculum suum*), as the most noble seat in the whole sensible world. Then where it says that he 'comes out of his pavilion as a bridegroom, exulting like a hero to run his race', I should understand this as referring to the radiating sun, namely to the light and the already mentioned calorific spirit, which fertilizes all corporeal substances and which, starting from the solar body, spreads very fast throughout the world; all the words correspond exactly to this meaning…Nor should I be silent about another consideration of mine, which is not irrelevant to the present purpose. I have already discovered the constant generation on the solar body of some dark substances, which appear to the eye as very black spots and then are consumed and dissolved; and I have discussed how they could perhaps be regarded as part of the nourishment (or perhaps its excrements) which some ancient philosophers thought the sun needed for its sustenance. By constantly observing these dark substances, I have demonstrated how the solar body necessarily turns on itself, and I have also speculated how reasonable it is to believe that the motions of the planets around the sun depend on such turning…

---

6 It is astonishing to realize how, when students in the classroom are exposed to the wondrous intricacies of the history and philosophy of science, they naturally tend to reawaken with surprise and satisfaction from the slumber of consciousness induced by the science classes. The most frequent phrase I receive as a comment from undergraduates, very often unsolicited, is the sigh of relief "…we understand!".

> Therefore, after singing praises to the sun…he goes on to the greater prerogatives of the divine law, and he wants to give it priority over the sun by adding, 'the law of the Lord is unspotted, converting souls, etc.'; this is as if he wants to say that the law is as much greater than the sun itself, as being spotless and having the power of making souls turn around itself is a more lofty condition than being covered with spots (as the sun is) and making corporeal and worldly globes turn around itself.[7]

Let us recall that Galileo proposed a hermeneutic principle, a first principle, for the interpretation of the Bible, according to which two truths cannot contradict each other. But the principle cannot be demonstrated though it can be illustrated by experimenting with the consequences that *ingenium* can draw from its application. Thus, Galileo participates in a humanist tradition that was fully conscious of the limitations of formally deductive science. Hermeneutics glimpses upon the nature of first principles, attempting to overcome the vicious circle of formally deducing the principles from other principles. The very word 'hermeneutic' comes from *hermetic* and cognate words. The exploration of the mysterious nature of the sun by both astronomical observation and biblical hermeneutics, including metaphysical and theological speculation, is part and parcel with Galileo's hermetic mode of being-in-the-world. It is a mode of existence in the life-world that need not be committed to a specific doctrinal content, and which resists the separation of self and world. Renaissance hermeticism teaches tonalities and performance, as we have seen, and a pragmatic approach towards understanding that can be adapted to different philosophical worldviews. Above all it teaches the divinity of the cosmos, the ecology of stewardship, and the participation of the human being in the unfinished project that God undertook with the creation of the world. There is a beautiful example of this performative hermeticism in Marsilio Ficino's *De sole*.[8]

Ficino invites the hermetic disciple to imagine the sun in the wake of Averroes who had first deprived it of matter. Then, imagine taking away from the sun all finite quantity, while leaving him his light and virtue, so that the light will remain there, infused with a power that could not be delimited by any quantity or figure. By this expansion the sun will become omnipresent to the imagination, engulfing the space in its entirety. As the sun exceeds the intellect so it surpasses the force of penetration of human vision. As God placed his tabernacle in the sun, you will be able to discover God according to the measure of your own forces. Further on, in the twin essay *De lumine*,

7 Finocchiaro 1989, pp. 65–66, slightly altered.

8 "On the sun". Cf. Ficino 1561, I, p. 971.

Ficino introduces the idea that light is proper to the circular body that moves by returning to the same position, and indeed to intelligence that returns to itself, and which, we might paraphrase, becomes self-consciousness.[9] We find buried in the Neoplatonic culture of Florence another clue to Galileo's reverence for circularity and to his abhorrence of the elliptical. The circle is not only an image of the divine and of the sun, but circular motion is the human intellect itself. The light of the sun will flood the sphere of the imagination of the disciple who explores the inner vision of the tabernacle of God. Before imagining turning a telescope to the sun, the disciple must have imagined the light of the sun, free of matter and figure and quantity, flooding the eye of the mind. Removing matter, figure, and quantity from the sun in the boundless space of the imagination means denuding the sun. The nakedness of the sun-God mates with the nakedness of the human imagination. Though the latter is flawed because humans have been cast in opaque material bodies, the light of the sun is penetrable by the imagination, though not as far as the center of the solar body where the tabernacle hosts the divine presence. The sunspots that can be seen by projecting images of the sun through the lenses of the telescope on white paper screens are also projections of the flaws that the opaqueness of the human body opposes to the shining light of the sun-God. The naked self does not see the nakedness of its divine center but only the shadows of the flaws projected outwards. The celebration of shiny nakedness in Galileo's poem was not a satire of the cumbersome toga but a celebration of nakedness as the opaque state of being before God.

In the controversy over the nature of sunspots with Christoph Scheiner, Galileo employs the language of scholastic metaphysics to defend the reality of the dark blemishes and their close proximity to, or position on, the surface of the sun (in contrast to Scheiner, who thought that they were stars, or clusters of starlets, orbiting the sun). The word *accident* recurs obsessively throughout the *Letters on sunspots*. Indeed, both accidents and substance are real being (*ens reale*), hence Galileo had recourse to scholastic language to reject the opinion that the incorruptibility of the heavens would be challenged by the sunspots unless the latter were explained away as unreal, that is, as hallucinations or deceptive phenomena due to the instrument. He famously proclaimed that the investigation of substances is a vain enterprise, and that accidents are the proper subject of scientific knowledge, for knowing the substance of anything is possible only in the state of grace that is reserved to the blessed.[10]

9 "On light". Cf. Ficino 1561, I, p. 983.

10 Galilei 1890–1909, V, pp. 187–188.

We cannot know the essence of the sunspots any more than the essence of clouds. But we can reason by analogy with the accidents of terrestrial clouds, such as their changeability, relative thickness, mobility, and fluid matter, and draw some conclusions. However, Galileo does not content himself with bric-a-brac similitudes, for he knows that the problem with the terrestrial analogy is that the dark matter of the earth does not shine but receives its illumination from the sun. How would the earth look like seen from far above its surface? He projects his imagination into the vast celestial spaces. This is the hermetic flight. He contemplates the appearances of the terrestrial clouds from on high. He sees the earth uniformly flooded by the light of the sun, and thus not showing any phenomena like the dark blemishes spotted on the sun. But if the earth were as luminous as the sun, the hermetic astronaut would see the bright planet punctuated by the dark spots caused by the interfering clouds.[11] The whole structure of Galileo's analogical argument in the *Letters* ultimately depends on the cosmic flight of the hermetic imagination.

The Solar Geometry of the Intellect

Christoph Scheiner's *Accuratior disquisitio* (1612) on sunspots, which Galileo examined and rebutted in the *Letters*, opened with a proof that lent itself to being recast in the circular geometry of mind. It was the perfect mathematical input necessary to activate Galileo's solar intellect.[12] The first part of the *Accuratior disquisitio* concerns the computations deemed necessary by Scheiner in order to show that Venus orbits the sun. Scheiner's argument is based on Venus' apparent path on the surface of the sun at inferior conjunction. The entire computation depends on arithmetic and, more precisely, on the application of the *regula aurea* (golden rule).[13] Its goal is to calculate distance QR (Venus's path on the surface of the sun). With reference to figure 4.1, let, Scheiner asks, the great disk represent the sun and the smaller disks represent Venus at different positions along the path towards the sun (IF). He now supposes that ten days before conjunction Venus is at I and that at conjunction it is at K. He

11 Galilei 1890–1909, V, pp. 106–107.

12 Cf. Scheiner's work in Galilei 1890–1909, V, pp. 37–70. The complete title is *De maculis solaribus et stellis circa Iovem errantibus accuratior disquisitio*.

13 The *regula aurea*, or golden rule, is probably the most important tool of Renaissance practical arithmetic. It basically allows one to calculate a fourth quantity given three other quantities forming a proportion. In modern algebraic terms, it is the solution of a simple equation.

then assumes that IL and LK are known and asserts that IK can be calculated by means of Pythagoras' theorem applied to the right-angled triangle KLI. By subsequently applying the *regula aurea* to the three quantities KL, LI, MN, he can calculate the fourth quantity NK. In the same way, he calculates MK given the three quantities KL, KI, MN. Scheiner now considers right-angled triangles MPI and EOM and applies a Lemma proven at the beginning of the *Accuratior disquisitio*. He shows that triangles MPI and EOM have two angles equal to one another, thus concluding that their sides are proportional.[14]

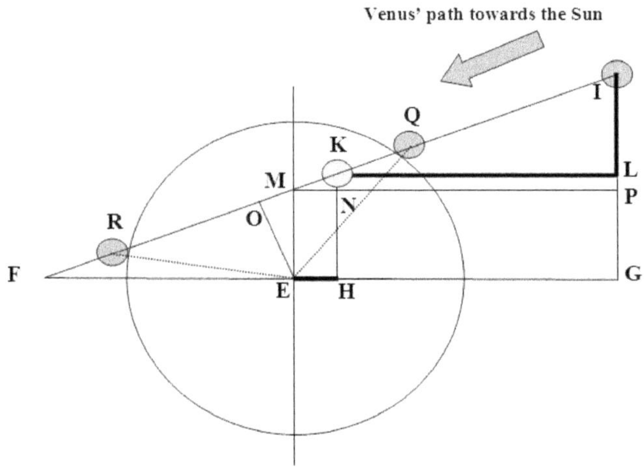

Fig. 4.1 Scheiner's diagram for the computation of the transit of Venus (adapted from Galilei 1890–1909, V, p. 42). The thick lines represent the known quantities assumed by Scheiner at the beginning of the proof.

Now, since all quantities in triangle MPI can be calculated from the data already known, by applying three times the *regula aurea* Scheiner can calculate all three sides of triangle EMO. Finally, by considering triangles EQO and EOR and applying once again Pythagoras's theorem, he determines distance QR.

In his *Third Letter*, Galileo begins by examining Scheiner's mathematical ability. Marveling at the long-windedness of Scheiner's proof, he says that he

14 I have slightly simplified Scheiner's reasoning for simplicity. Although the proof of the *Lemma* is indirectly revealing of Scheiner's arithmetical mathematical style, its details are irrelevant for our purposes. Cf. the *Lemma* in Galilei 1890–1909, V, pp. 40ff.

would have liked him to be a more proficient geometer.[15] But what is clear from the tone of the passage is that Galileo is not really interested in mocking Scheiner's mathematical 'clumsiness'. He simply notes that the whole proof has the objective of finding distance QR, given GH, HE, KH, and IG, and that it could have proceed by means of proportional reasoning as follows. Since KH, IG are known, IL and LG can be found. Thus, since as IL is to LG so IK is to KF, and GH is to HF, and since IL, LG, and GH are known, then HF becomes known. In addition, since triangles FEM and KLI are similar, as FE is to EM so KL is to LI. But KL, LI, and FE are known, therefore EM becomes known. KI can be calculated since in right-angled triangle KLI the sides KL, LI are known. Thus, since as IK is to KL so ME is to EO (because triangles KLI, MEO are similar to triangle FEM and therefore similar to each other) and IK, KL, and ME are known, EO can be found. Finally, since ER is given, RO and RQ can be calculated.[16] This is all that Scheiner would have needed to calculate Venus's transit, says Galileo.

Galileo's solar intellect played an important role in debate and controversy. It is the presence of the solar geometry of proportions that explains Galileo's almost 'automatic' geometrical rendering of Scheiner's arithmetical analysis of the transit of Venus. Galileo transforms Scheiner's arithmetic-based schema into a geometrical diagram by recognizing the similarity of two triangles (LIK, KHF). Note that (a) since Euclidean plane similarity reduces to proportionality through the *Elements*'s Proposition 1, Book VI, and (b) since the construction of Proposition 1, Book VI, is based on congruence criteria for plane figures (i.e., on visual criteria of equivalence), Galileo's solar reasoning about the transit of Venus returns on itself like a circular motion. Indeed, the imagination of the transit cannot be 'computed' arithmetically. In other words, although Galileo can easily construct a linguistic proof, he cannot do so without relying on the congruence criteria underlying Euclid's theory of plane similarity.[17] Congruence reasoning returns on itself by recognizing the

15 'Ma più presto avrei desiderato Apelle [Scheiner's pseudonym] alquanto più resoluto geometra…'. Galilei 1890–1909, V, p. 193.

16 Galilei 1890–1909, V, p. 194.

17 Galileo's proportional reasoning is a form of reasoning based on the principled manipulation of ratios and proportions, according to the rules set forth in the fifth book of Euclid's *Elements*. On Galileo's use of proportional reasoning in natural philosophy, a considerable body of literature is now available, which allows us to understand most of its technical aspects better. One should keep in mind that a Euclidean 'ratio' is not a fraction. It is defined by Euclid as a sort of 'relation'. In Renaissance Latin editions of the *Elements*, 'habitudo' is the typical word used to define ratio. Cf. Armijo 2001, Drake 1973, 1974, 1987

sameness of the quantities that are placed against one another. There is something astonishing in Galileo's reaction to Scheiner's arithmetical proof. It is not that he needed to amend his opponent's argument, since Scheiner was only cumbersome but not wrong. He really needed to reassert the primacy of solar reasoning. It was more than a penchant for impeccable mathematics. It was almost like his natural mode of being. The sun was a modality of being-in-the-world that we cannot separate from Galileo's self.

Heliocentrism and Cardiocentrism

Galileo's argument for heliocentrism based on his exposition of biblical passages includes the powerful analogy of the sun and the heart of the animal. The heart continuously regenerates the vital spirits that sustain life and vivify the body while at the same time receiving the nourishment that is necessary for its functioning from the body. Likewise the sun vivifies the celestial bodies that rotate around it while at the same time receiving light from the stars that fortifies its own inner light.[18] Galileo's heliocentrism is physiological. It is the imagination that the sun-God vivifies the organs of the cosmic animal. To defamiliarize ourselves with the received image of heliocentric astronomy that has been handed down to us by the historiography of the scientific revolution, we need to restore the physiological significance of Galileo's analogy. We need in effect a historical phenomenology of the heart. My discussion will be based on the defense of the centrality of the heart in the constitution of the animal substance by Cesare Cremonini.[19] His philosophical magnum opus, based on more than three decades of teaching at Padua, *On the origin and primacy of the members against Galen*, was the culmination of cardiocentric thought in Western culture before the scientific revolution of the seventeenth century irreversibly shifted the focus of philosophers and scientists in Europe towards the brain. It is not so much the detailed theories that are relevant here as the alien mode of thinking that depends on the centrality of the heart as opposed to our familiarity with the dualism of mind and brain.

Before the advent of the brain absolutism over the last four centuries, the heart is the principle organ of animal life, the embodiment of a notion

(reprinted in Drake 1999, II, pp. 265–280, II, pp. 292–306, III, pp. 61–75, respectively), Frajese 1964, Giusti 1981, 1986, 1990, 1992, 1993, 1994, 1995, Palladino 1991, and Palmieri 2001.

18 Galilei 1890–1909, p. 302.

19 Cremonini 1627.

of flesh that transcends the psychophysical worldview of the dualism of mind and brain. The cardinality of the heart in structuring organismic sentience is genetically (in the order of spatial-temporal nature) and ontologically (in the order of being) absolutely primordial. The heart is elemental flesh in this sense. An inanimate element such as fire is corporeal substance resulting from an inner causality, or dynamism, which is the formative tension of contrary qualities such as hot and cold, dry and moist, resisting each other in the struggle for bringing about the corporeal, individuated existence of the element within a finite region of space and time. The generation of multiform living substances, such as human animals, that are subject to generation and corruption, i.e., that do not exist coeternally with being, is necessarily and temporally a progressive manifestation of regulated self-growth of mutually interdependent member-substances from one cardinal member-substance, which is the elemental flesh of the heart. It is a process of gradual addition of heterogeneous parts accruing from the struggle of opposites (occurring at the more fundamental levels of the inner dynamisms of the single member-substance) that occupies a finite duration, and which must begin from a unit of origination at a specific point in time. The generation of accidental forms such as color is rather different insofar as their process of generation might be instantaneous, because the supervening of a quality in a subiectum need not occur as the progressive striving of opposites in tension that is the inner dynamism of substances. However, the dynamic polarization that establishes the appearances of substance and accident is reversible insofar as we know that the conflate model relates the domains of substance and accident dynamically.

Thus, the process of generation and corruption of substances and the process of possibly instantaneous alteration of qualities supervene on the underlying dynamic polarization that determines the manifestation of substantial being as opposed to accidental being. The manifestation of substance such as element as opposed to quality such as color is always the result of a dynamic stratigraphy that directs the flow of being along trajectories in continual transformation. The flesh of the heart is this privileged member-substance that is the originary starting unit of the whole process of growth of an animal substance. The embryological evidence by which the heart must be assumed to be this originary starting unit in higher animals need not concern us here, but one philosophical mediation can be proposed that intimates at the primacy of the flesh of the heart. The flesh of the heart is this originary starting unit because, in order for it to be the first member-substance that is wholly self-sustaining as the autonomous living process that will harmonize the growth of diverse member-substances, its corporeal structure must include the totality of the structures of the parts of the organism that will be orderly constructed as if they were already implicated within its own originary corporeal makeup. This totality of structures already implicated in the corporeal makeup of the flesh of the heart is involuted in its own absolute self-implication. The problem with thinking this absolute non-spatial and non-temporal self-implication of the originary heart appears to me to be due to the imprisoning of the modern mind within the confines of mechanistic modalities of thought. These were determined by the classical vision of particles colliding into each other in order to form assemblages. Unlike fundamental physics which in the twentieth century has prosecuted its own classical foundations, biology has not yet grown out of this outmoded way of thinking. Corpuscles are not the fundamental units of matter. Living bodies are not assemblages of subatomic particles forming larger biological entities such as cells and tissues. The flesh of the heart is the most originary manifestation of matter in the corporeal universe. This manifestation of matter is temperament in its originary enfoldment since the fabric of the heart is compounded of a diversity of structures that is this primordial implication of all the structures to be deployed in the animal body. In the living animal soul is the substance informing the flesh of the heart. This temperament, therefore, is not only a temperament of qualities such as can be found in inanimate substances, but is a living temperament. Life is primordial vital heat that is most orderly tempered in the flesh of the heart of the animal. The brain is frigid, insisted the Padua Averroistic philosophers to the European students caught in between Aristotle and Galen. It is the heart, the seat of *calidum innatum*, innate heat, that must be restored to primacy, right at the center of the spheres of animal sentience, for thought to awaken from the delusions of the brain. Frigidity in the brain, which is its constitution, its temperament, means barrenness in thought, this was the understatement to win over the perplex students.

I return to Galileo's image of the heart-sun. The solar thought that instated the sun at the center of the universe was physiological through and through. The great master of libertine Averroism at Padua was also the great defender of the centrality of the heart. While wearing the mask of Aristotle, the unbeliever Cremonini was as much 'Copernican' in physiology as the hermetic Galileo was in astronomy. But the difference that blocks our understanding of both is that one has been elevated to the rank of a hero or martyr of science, while the other has been forgotten as the villain of the plot. Both have become unintelligible for opposite reasons. But there is more. The God-heart-sun of Galileo exalted the androgynous body of the cosmic animal. The intolerant differentiation of the sexes necessitated by Christian ideology and institutions such as the university has done us a certain amount of damage. Restoring the thought of the heart opens the way to restoring images of the androgynous body that were prevalent before modernity set the course of science towards the sexed brain and its dualistic pathologies. Galileo and Cremonini were cardiocentric queers hiding behind Copernican and Aristotelian masks.

The Sun and the One

Galileo extensively quotes Pseudo-Dionysius the Areopagite (*De divinis nominibus*) in support of his biblical exegesis. He wishes to fortify his belief in the luminous and vivifying spirit emanating from the sun.[20] But he goes beyond scriptural hermeneutics and stresses that Dionysius's philosophical emphasis on the force of the sun is the royal road to philosophy. Dionysius's mystical philosophy of the one not only conspires with Galileo's solar intellect but pervades his fascination with the paradoxes concerning the one and infinity. In the passage quoted by Galileo, Dionysius speaks of the sun as one.[21] "For, if our sun, at the same time that he is one and sheds a uniform light, renews the essences and qualities of sensible creatures, although they are many and various, and nourishes and guards, and perfects and distinguishes, and unites, and fosters, and makes to be productive, and increases, and transforms, and establishes, and makes to grow, and awakens, and gives life to all; and each of the whole, in a manner appropriate to itself, participates in the same and one sun; and the one sun anticipated in himself, uniformly, the causes of the many participants; much more with regard to the Cause of it and of all things, ought we to concede that It first presides over, as beseems One super-essential Oneness, all the exemplars of things existing…".[22] If we extend the horizon of interpretation by reading the chapters of *De divinis nominibus* that follow the excerpt cited by Galileo, we find an extraordinary discourse, in the conclusive section, concerning both the one and perfection, as attributes of God, in relation to numbers. The places in Galileo's *Discorsi* where he launches into speculative

20 "Ora, stante questa filosofica posizione, la quale è forse una delle principali porte per cui si entri nella contemplazione della natura…" (Galilei 1890–1909, V, p. 303).

21 Dionysius 1566, p. 53 recto.

22 Dionysius 1897, p. 81.

contemplations of the paradoxes of infinity betrays his being captivated with the mysticism of Dionysius. Obviously, Galileo's formulation of the paradox that odd and even numbers, and square roots and squared numbers appear to be as many as all natural numbers was not directly borrowed from Dionysius. Galileo points out that we cannot predicate sameness, greater than, smaller than, of infinites in the same way as we predicate them of finite numbers.[23] However, he has learned how to think about the infinite progression of numbers by immersing himself in a life-world in which biblical hermeneutics became entangled with the mysticism of Dionysius. Galileo speaks of the '… discorrer che noi facciamo col nostro intelletto finito intorno agl' infiniti' (… the discoursing that we make with out finite intellect around infinities). This 'discorrer' of the intellect around the infinite is a running around but not necessarily in circles. The infinite cannot be grasped by this running around of the intellect. Even though Ficino introduced the idea that the light of intelligence returns to itself circularly, the finite intellect seems to be unable to revolve around the infinite because it is constantly falling into the trap of transposing to the infinite the ordinary attributions of comparison that are valid for finite numbers. The infinite exceeds the possible circles of the intellect because it cannot be confined in the finite domain of shapes. The running around that vainly chases the infinite is the guiding idea that Galileo seems to be blindly or instinctually following in his search for enlightenment. The image of running evokes also the idea of the running away from something unbearable. It is thus an image of the anxiety that paradox elicits because its incomprehensible perfection escapes the limits of the finite intellect.

Dionysius says that the perfect is not only that which is perfect in itself, and which is defined uniformly by itself and uniquely, but that which overcomes perfection insofar as it antecedes everything, determines all infinites, and explicates itself beyond all confines; and, furthermore, it is not comprehended or captured by anything, but attains everything and beyond everything by perpetual progression and boundless activity. Furthermore, perfect also is said to be that which is never increased, and which is always absolved from its own numbers, and which cannot be diminished, in that it has everything in itself, as if anticipating everything, and is redundant and endowed with a

23 "Queste son di quelle difficoltà che derivano dal discorrer che noi facciamo col nostro intelletto finito intorno agl' infiniti, dandogli quegli attributi che noi diamo alle cose finite e terminate; il che penso che sia incoveniente, perchè stimo che questi attributi di maggioranza, minorità ed egualità non convenghino a gl' infiniti, de i quali non si può dire, uno esser maggiore o minore o eguale all' altro" (Galilei 1890–1909, VIII, pp. 77–78).

unique and perpetual and cumulative dispensation that is perfect and perfects everything. But it is said one because in a singular and excellent mode is the cause of everything without being assimilated into anything. Thus, everything and all particles of all things have the one in themselves, and everything which is, it is in virtue of this one. But this one is not the unity of a multitude, but is prior to units because it determines all units and multitudes. For, a multitude cannot be without the one, though the one can be without a multitude, in the same way in which the monad antecedes all numbers determined by multiplication and increase. The one is the element of all things. Thus, if the one is taken away, there will remain neither the whole, nor the part, nor anything of all there is.[24] There is in every paradox a potential for deception. It is the constant temptation to grasp the paradox by running around it. The infinite number, Galileo argues, seems to be beyond all finite numbers. But if we keep counting, can we really hope to reach infinity, in the sense that we are really attaining a better and better approximation of the infinite number? It would seem so, as a million surely must be closer to infinity than one hundred or one, and yet this is also highly misleading. As we consider larger and larger numbers we will find fewer and fewer squares and cubes (for instance, there are ten square numbers in one hundred, but only one hundred squares in ten thousand). Hence, says Galileo, it seems that, since it was concluded that in an infinite number there should be as many integers as squares and cubes, and so on, it is rather the case that the larger the number the farther removed it is from the infinite number. Conversely, by returning towards the one, the 'density' of the squares, cubes, etc., increases. Thus, if any number may be said to be truly infinite, it must be the one. For, in the one there are all conditions and requisites of the infinite number, namely, its containing in itself as many squares and cubes as numbers. The one is in fact the square, the cube, and all powers (insofar as it remains the same one no matter how many times you multiply it by itself), and hence there is no particular property of the powers that cannot be found in the one. It must be conceded, then, that there is no other infinite number but the one.[25]

I am not so familiar as I would like to be with the cultural and theological context of Dionysius's mysticism of the one. I also think that the label 'mysticism' is suspicious, and that we should do well not to compare and contrast 'mysticism' with the equally suspicious 'reason' or 'rationalism'. But Galileo's *Discorsi* is replete with discourses that transcend the domain of the finite.

24 Dionysius 1566, p. 61 verso–62 recto.

25 Galilei 1890–1909, VIII, pp. 82–83.

For instance, there is the notion of the degree of instantaneous speed (*gradus velocitatis*), which has become part and parcel with the language of the calculus that was developed in Newtonian classical mechanics. The science of uniformly accelerated motion begins informally with a discourse about *gradus velocitatis*. It appears to be paradoxical, says Galileo, that a body starting from rest will pass through infinite degrees of speed, with infinitely small variations, as we normally see falling bodies immediately coming down with great speed.[26] The paradox requires the finite intellect to think through the accelerating motion as the infinite transition of a material body through infinitely many and infinitely small variations of speed. Galileo's mechanics rests on this paradox, as it could not dispense with the notion of the instantaneous degree of speed. In medieval scholasticism speed was a quality of the moving body, and its variation always remained a difficult question in scholastic natural philosophy. But Galileo's mechanics requires the instantaneous degree of speed as absolutely foundational. His new science of uniformly accelerated motion cannot take off the ground without accepting into its deductive structure the paradox of the infinite degrees of speed, each of which must be instantaneous so that the body does not persist in any degree for a finite interval of time. On this condition the finite intellect understands that an infinity of degrees is actually possible because the infinite transitions do not add up to form an infinite duration. It is not an exaggeration, I think, to suggest that biblical hermeneutics and the paradoxes of infinity and of the divine perfection of the one are entangled. They constitute the life-world in which Galileo's solar imagination can soar. They are as essential as the venerable geometry of his 'divine' Archimedes. In fact they are not separable from Galileo's self. They constitute its powers beyond Galileo's conscious identity.

It is mysterious how mathematicians succeed in discovering their demonstrations. The delusion that mathematics is exact because it carries an intrinsic force is due to its deductive style of presentation. But deduction is not how proofs are discovered, for at any single phase in a deductive process there are open infinite possibilities to move forward in the process and the secret is the 'reason' that steers the course of the process. It is easy to convince oneself that

26 Galilei 1890–1909, VIII, p. 199. Galileo reports that he had this difficulty initially, but that he was able to overcome his doubt with the analogy of the force of percussion. He also adds the argument that a body thrown upwards would decelerate up to the point of rest before inverting its motion and that there is no reason to think that it would not go through infinitely many degrees of decreasing speed up to the degree zero, and hence that the same must happen to the body falling from rest.

this is the case by trying to demonstrate the Pythagorean theorem following the Euclidean postulates and the preliminary theorems proved by Euclid up to the point where he can furnish the celebrated proof. Very few people will succeed. Superficially one might conclude that Galileo's strategies for problem-finding and problem-solving were the rational application of geometry to physics. But from the perspective of the creativity processes that fuelled his hermetic imagination the vista is quite different. We might speak of emergence, which is a relatively new subject in the understanding of how mathematics develops and changes over time. The proliferation of centers of interest, before the commitment is set to a final form of the problem, lies at the core of the dynamics of emergence. This peculiar role of unconscious emergence has long been recognized by scientists. Almost a century ago, in a fascinating conference on mathematical creativity, Henri Poincaré emphasized the role played by the unconscious in sparking mathematical insights.[27] Poincaré's language has the hallmarks of the aesthetic feeling that accompanies the final slotting into place of all the pieces of a solution to a problem, the glimpse of a new emerging order. Poincaré talked of 'sentiment', of the "ordre mathématique, qui nous fait deviner des harmonies et the relations cachées".[28] Further, Poincaré reveals that the solution to a question concerning Fuchsian functions came during a night of insomnia caused by black coffee, where "les idées surgissaient en foule…". In the morning all that remained to do was to write down the results, which, Poincaré says, he did in a few hours.[29] At another place, he says that "la solution de la difficulté qui m' avait arrêté m'apparut tout à coup".[30] He claims that "le rôle de ce travail inconscient, dans l' invention mathématique, mais paraît incontestable…".[31] He further highlights the alternating periods of conscious and unconscious mental activity as being central to his experience of mathematical invention. Feeling and a sense of elegance are by no means alien to the achievement of mathematical proof. "On peut s' étonner de voir invoquer la sensibilité à propos de démonstrations mathématiques, qui, semble-t-il, ne peuvent intéresser que l' intelligence. Ce serait oublier le sentiment de la beauté mathématique, de l' harmonie des nombres et des formes, de l' élégance géométrique. C'est un véritable sentiment esthétique

27 Poincaré 1908.
28 Poincaré 1908, p. 522.
29 Poincaré 1908, p. 523.
30 Poincaré 1908, p. 523.
31 Poincaré 1908, p. 524.

que tous les vrais mathématiciens connaissent".[32] One thing is clear, I think. We need to enrich our histories and philosophies of science. We need to study unconscious emergence and the imagination, the processes behind the deductive style of presentation, and the mathematician's aesthetic embracement of the life-world. Galileo's life-world, which cannot be separated from Galileo's self, was incredibly rich in terms of cultural, philosophical, theological, and aesthetic interests, and the divergent directions of his pathways took him towards marginal and/or sensitive areas of the culture where he found himself alongside primordial moods, conjuring fantasies which he would not otherwise have dared to entertain.

The Pleasure of *Anima Mundi*

On 15th January 1633, as he was about to leave for Rome, summoned by the Holy Office, Galileo wrote a letter to his friends, Elia Diodati and Pierre Gassendi, lamenting that his *Dialogo* was most likely going to be prohibited by the Roman authorities. However, the letter is interesting for another reason. Galileo acknowledged that he had read Gassendi's *Epistolica exercitatio* against Robert Fludd's philosophy with special pleasure.[33] Gassendi had written the *Exercitatio* at Marin Mersenne's invitation, in order to discuss and rebut Fludd's response to his own attack on Fludd's philosophical views. The intricacies of the controversy are of little interest for my immediate purposes here (though much interesting per se). The truly fascinating question concerns Galileo's pleasure. What did Galileo find that was so much to his taste in Gassendi's writing? Not (or not only) the astronomical appendix that also Galileo mentions. The letter's language is unequivocal. His pleasure was about the *Exercitatio*. Why? I will not answer this impossible question, for, alas!, Galileo

32 Poincaré 1908, p. 525. With memorable words, which, I believe, reflected his own mathematical practice, Évariste Galois had already expressed an analogous view. "Du fait bien constaté que les efforts des géomètres les plus avancés ont pour objet l' élégance, on peut donc conclure avec certitude que qu'il devient de plus en plus nécessaire d' embrasser plusieurs opérations à la fois, parce que l' esprit n'a plus le temps de s' arrêter aux détails" (Galois 1962, p. 9).

33 "Ho letto con particolar gusto l' Esercitazione del S.r Pietro Gassendo contro alla Fluddiana filosofia, come anco l' Appendice delle osservazioni celesti" (Galilei 1890–1909, XV, p. 26).

tells us no more.[34] But I will explore the *Exercitatio* in search of motives that resonate with hermeticism and hermeneutics, with a view to *anima mundi* and Galileo's biblical exegesis of the subtle spirit emanating from the sun. My focus is not doctrinal. Mersenne's, Fludd's, and Galileo's doctrines were worlds apart. First, I am interested in Galileo's pleasure as a tonality structurally present in his life-world, that is, prior to the polarization of experience and to the argumentative violence forced by early modern theological-philosophical controversialism. Second, I am interested in the possibilities of understanding projected by the pleasure of being-in the-world of hermeticism. The persecutory, arrogant, and aggressive tone of Mersenne's assault, a typical instance of the *odium theologicum* (theological hatred) that erupted in Europe in the aftermath of the religious reforms, was candidly repudiated by his friend Gassendi, who moderated between his fellow Frenchman and Fludd. Gassendi made no bones about it. It had been Mersenne who had started the fight, and frankly, he added, one could say that Fludd's reply, so zealously deprecated by Mersenne, was not much worse than the latter's vitriolic attack, which went so far as to call upon the political authorities to punish the 'stinky magician', the 'heretical Fludd'.[35]

---

34 There is another intriguing instance, again, revealed in a letter, in which Galileo expands on pleasure as a modality of learning, but it is in a different context, the experimental work with armed magnets. "I have been for three month caught with an admirable affair, that is, the artful multiplication of the lodestone's virtue to hold iron. I have already been able to arrange things so that a 6-ounce lodestone, which by its own virtue holds no more than one ounce of iron, can with art hold 150 ounces, but I hope to improve. I will report to You since You have a speculative mind very much inclined to delight in this sort of accidents. As for myself I cannot marvel enough at all of this, when I see such an angry conjunction occurring because of a simply immaterial virtue. I am so much more pleased with it as I know that Gilbert, who speculated deeply about this, and experimented and wrote diligently about it, could not increase the virtue of a lodestone holding one ounce to holding more than three [...]. The gain, which I have been making day by day gradually, has ensnared me with so much pleasure and marvel that I have almost become a locksmith, and being intrigued by this I have shut away everything else; to the point that, becoming more and more insatiable, like a miser, I cannot be satisfied. At the beginning I thought that it was a huge profit to increase the lodestone's virtue forty times, but now the interest rate of one hundred and fifty times is not enough, and I am at pains to improve as much as I can, while at the same time I am learning what it means to be a miser" (Galilei 1890–1909, XIII, p. 328).

35 "Sane vero hic, mi Mersenne, negare non licet, quin tu ipsi scribendi ansam aliquam feceris. Revera enim dici potes paulo acrius illum tetigisse. Ac zelus quidem, quo evectus es, commendari cum debeat, attamen te latere non potest quin admodum durum sit viventi in Christiano orbe appellari *cacomagum*, *haeretico-magum*, *seu foetidae*, *et horrendae magiae*

It is strange that the moderate tone used by Gassendi, who, although he did not have much sympathy for Fludd's philosophy, had no qualms either about reprimanding his good friend Mersenne, has been downplayed in the literature. An effort is needed to defamiliarize ourselves from the positivist totalitarianism that has repeatedly denigrated the naturalistic, magical, and hermetic worldviews that preceded and accompanied the rise of the early modern sciences. For Francis Yates, "Mersenne is a modern; he has crossed the watershed and is on the same side of it as we are; belief in the power of magic images of the stars seems to him quite mad. [...] Mersenne, who completely discards astrology, naturally also discards astral magic, the miraculous virtue of plants, stones, images, and the whole apparatus upon which *magia naturalis* rested. [...] In the course of this energetic clearance, which demolishes the Hermetist-Cabalist basis of Renaissance magic, Mersenne mentions and condemns all the chief propagators of these views [...]. The life-blood of the Renaissance Magus drains away under this onslaught; his most cherished theories, illusions, and delusions are turning into so much useless lumber in the cold clear light of the new age".[36] This picture painted by Yates (who infelicitously, though certainly unwittingly, uses the same word 'propagator' used by Mersenne to vituperate Fludd) stands in the way of understanding. It is an embarrassing legacy of the theological and philosophical hatred that we 'moderns' inherited from the religious wars of the sixteenth and seventeenth centuries. We should do well to pause and ponder the genocidal and ecological catastrophes that our crossing the watershed has visited upon the world since the miraculous advent of the cold clear light of the new age. How many times in our history has the energetic clearance of the views of the 'other', ideologically, religiously, racially, and/or scientifically motivated, turned into the ethnic cleansing of the 'other'? In my life-world the words 'clearance' and 'cleansing' resonate eerily, echoing from the Middle East through the Balkans to North America. The intolerance for the 'useless lumber' that we left behind, concomitantly with the exaltation of Christianity as the true religion

---

doctorem, et propagatorem: audire, *non esse ferendum huiusmodi doctorem impune*; provocato principe, *uti de illo poenas sumat*, minisque etiam adhibitis, eundem propterea *brevi submergendum fluctibus aeternis*, etc. Ut nihil dicam de atheismo, atque haeresi, quam tu quoque obijcis Fluddo" (Gassendi 1630, p. 9).

36 Yates 2002 (1st ed. 1964), pp. 474–476. For a well-calibrated evaluation of Gassendi's role in the Fludd-Mersenne-Gassendi controversy, see Clericuzio 2000, pp. 71 ff. See Rösche 2008, for an excellent monograph doing justice to Fludd's hermetic philosophical system.

and of its 'secular' alter ego, Western science, gave ourselves permission to exterminate ethnic groups, cultures and civilizations.

Gassendi mounted a fascinating and subtle operation which, I would argue, shows intelligent respect for Fludd's useless lumber. About one third of the *Exercitatio* is neither apologetic nor polemical. It is a summary of Fludd's principles of philosophy judiciously excerpted from Fludd's works. It is a primer presented for the benefit of the reader, but also for the benefit of Gassendi's own understanding, which aims at genuinely penetrating Fludd's mind and hence his own mind.[37] This is my first highlight. Gassendi is showing the reader by his own example that non-violent being-in-the-world in the face of the alien 'other' can only be disclosed by a passivity of one's will to power in which one seeks the principles that guide the alien 'other' in their thinking. It is not prejudicial to one's thought to open one's mind to the principles of the 'other', on the contrary it is a requirement that precedes the possibility of genuine understanding of one's own mind. This openness of Gassendi's mind contrasts sharply with Mersenne's intolerant will to power, who must be considered the second author and/or the patron of the *Exercitatio*. He prefaced it with a long letter explaining and lamenting the occasion of the polemic, and crucially, with an expert opinion by a representative of the Sorbonne theological faculty that condemned Fludd's philosophy. The overall message of the *Exercitatio* as a polemical package against Fludd remains ambiguous because of its composite nature. It must be noted, however, that Galileo refers explicitly to the *Exercitatio* as authored by Gassendi when he speaks of his pleasure in reading it. He totally ignores the belligerent preface by Mersenne and the theological verdict. This comes as no surprise if we consider that in the letter to Diodati and Gassendi Galileo wrote a long excursus on his principles of biblical exegesis in which he reiterated that in disputes of natural philosophy one ought to start not from Holy Scriptures but from nature insofar as nature is the work of God and as such its authority should be considered on a par with that of Scripture. He was saying in between the lines that Mersenne was taking the wrong approach, and that in philosophy one should reason as a philosopher and not as a theologian, and consequently that it was inappropriate to invoke theological verdicts in order to condemn philosophical views no matter what Mersenne thought of Fludd. Of course the theologians would have countered to Galileo that if both the word and the work of God are equally authoritative, the question arises as to which should take precedence in settling cases of discordance, a legitimate question as both they and Galileo admitted that no

37 "Sic enim demum ipsius mentem venari perfacile erit" (Gassendi 1630, p. 11).

discordance was acceptable. Indeed, as we have seen, Galileo sought in biblical exegesis more than concordance. He projected his deepest possibilities of understanding the structure of the cosmos by making the sun the receptacle of divine spirit or *anima mundi*. It is clear that Galileo's emphasis in the letter falls on the question of natural-philosophical *controversy*. It is the latter that, he insisted, should not be regulated by concerns with biblical exegesis, yet his genuine modes of understanding, his tonalities and his innermost pleasures in discovering both the word and the work of God, were quite anther, *hermetic* matter.

Here is a brief summary of Gassendi's interpretation of Fludd's philosophy.[38] Gassendi begins by stating that Fludd posits two first principles of philosophy, light and darkness, the former being a universal agency, a formal principle of all things that pervades the whole universe by riding a most subtle spirit, and which insofar as it is pure and most real actuality causes all things to be and to be act. It is also called nature, essence, *anima mundi*, and God. By matter Fludd intends the material and passive principle by which all things are embroidered together. It has, according to Fludd, a quintuple nature, as he recognized by means of the chemical art, experimenting with wine and wheat. He first extracted from wine spirit, secondly an oil, thirdly he rectified a phlegm, fourthly another oil, and finally he purified the scum with ablutions. He mixed five equal portions of these substances in a hermetic vessel, sealed it and let them rest overnight, until the substances separated again occupying their proper regions, in order from the bottom, earth, water, air, fire, and a quintessential one he calls ether. An analogous chemical operation with wheat brings up the same partition of the mass into the five elemental substances. On this great experiment, says Gassensi, Fludd's entire philosophical system depends. The first principles of light and darkness can be considered before creation (when only God existed), during creation, and after creation, namely, in the created world. The first situation is a state of mind, or a state of reason, insofar as we consider these principles without any relation to the created things, from which, however, aside from this possibility of considering them as separated by this act of abstraction or mental severance, they are de facto never separated. Thus, they exist before creation as a radical unity from which the creation of all things depend. This light is said to be in darkness precisely because before creation it does not communicate with created beings. Fludd thinks it is therefore the hidden God. This light contracted in itself before creation is what the Cabalists call the non-being, or the infinite,

38 Gassendi 1630, pp. 13 ff.

in that it is communicable to infinite things but before creation it is involuted by infinite darkness. It is also said *smallness* in that as an originary unit it is root, square, and cube to itself. It is also said *eternity* in that it is outside of the temporal world and as such it is the archetypal world.

As to creation, Fludd considers it *creation from nothing* of any thing in the same sense as usually intended by theologians. But Gassendi warns that Fludd takes the Mosaic account of creation as a symbolic system. Hence matter or darkness is interpreted by Fludd as the nothing from which things are created by light, the hidden God. Thus when light emerges in created things it is also called the visible God, or brightness, or greatness in that it acquires the essence of multiplicity. Smallness erupts in creation to form innumerable things by means of greatness. There is a second sense in which, according to Gassendi, Fludd considers the work of creation. It is the alchemical operation. He starts from fermentation of wheat matter and proceeds to separate five substances that are given the names of elements. To this alchemical operation Fludd accommodates the mystery of creation. He thinks that the veiled account of Genesis hints at four realms of darkness, the first he calls *inanity*, the second *watery abyss*, the third *chaos* as confused conglomeration of the four elements, the fourth *earth*, namely, the excrement of the elements, in whose center he claims that the solar substance is contained. When the two principles of light and darkness are considered after creation, one recognizes that created things are mixtures of the principles in different proportions. But light is the principle or intimate essence that causes created things to operate, while darkness is the principle that distinguishes them and set them apart in prescribing their limits of action. Then, Gassendi congratulates Fludd on demonstrating experimentally that all things are made of atoms according to the philosophy of Epicurus (which makes us wonder what Mersenne would have made of this blunt concession to his rival). The fifth substance, namely, the ether is the principle of life that immediately receives in itself and broadcast the principle of light. Hence light penetrates first the ether, and subsequently after dressing itself in this ethereal garb pervades and vivifies all things. Hermetic philosophers call this ethereal substance a middle one in that it is neither corporeal nor incorporeal but rather a collagen uniting both. Furthermore, the two principles of light and darkness can be considered insofar as they enter into the constitution of the three worlds the archetypical world, the macrocosm, and the microcosm. By macrocosm, according to Gassendi, Fludd means the combinations and disposition of the five substances, the fifth of which includes in itself the principle of light. It is divided in three regions or heavens, the empyreal (almost entirely made of light), the ethereal (made of the remaining ether and light), and the elemental (made of the four elements). When light emerges into creation the first substance to capture it is the ethereal spirit, which is said *created light*. This created union of spirit and light is called by Fludd *anima mundi*. It animates all particles of matter since portions of *anima mundi* pervade all animated things.

In the final analysis, Gassendi admits, quite honestly, that he finds the discussion of *anima mundi* beyond his remit, and insofar as it may possibly touch on matters of faith he is ready to defer to the church.[39] If by *anima mundi* Fludd meant God, then Gassendi argues that his position appears to be compatible with Scripture. But since he clearly means something quite different, then it is the responsibility of theologians to dispute whether his position is ultimately tenable or not. Two main points in Fludd's conception of *anima mundi* appear to Gassendi to be theologically suspicious. First, Fludd wants God to enter the composition of the created world, both directly in the form of the ethereal spirit and indirectly by diffusion of spirit throughout the universe. Second, since he claims that the sun is the main receptacle of the ethereal spirit, namely, of *anima mundi*, he consequently claims that this diffusion is effected by the solar rays, and thus the animating particles of *anima mundi* are transmitted through the universe to all living things.[40] In sum, Fludd's concept of *anima mundi*, which might even resemble the abhorred universal intellect of the Averroistic interpretation of Aristotle, appears to Gassendi to be most likely untenable on theological grounds.

Gassendi's disputation with Fludd has come full circle. He has revisited Fludd's philosophical system starting from its empirical basis of distillation and fermentation of substances. He has applauded patterns of explanation of the physical make-up of the universe that in essence seem to be reducible to the atomism of Epicurus, and finally he has discussed the most delicate theory of *anima mundi* in the light of Fludd's first principles. He doubtlessly found it untenable on grounds of Christian theology. Gassendi underlined something important about Fludd's way of proceeding in natural philosophy. He emphasized that Fludd had started from experiment. He noticed that his philosophical system had an empirical basis and that Fludd was looking for points of convergence between experience and biblical exegesis. In other words, Gassendi's exposition makes Fludd look first and foremost like a natural philosopher intent on experimentation as the prime source of physical knowledge,

39 Gassendi 1630, p. 118.

40 Gassendi 1630, pp. 120–121.

and secondly like an interpreter of Scripture who is looking for illumination in the interpretation of the Bible, and who finds this illumination in the empirical knowledge gained from experiment. There is an uncanny parallel with Galileo, who used his discovery of the rotation of sunspots as a source of illumination for interpreting the Bible, pointing to the motion of the sun and to the sun itself as the receptacle of spirit. On the other hand, Gassendi privileges the Bible. He dismisses Fludd's hermetic theory of *anima mundi* on theological grounds. Both Fludd (or, shall we say, Gassendi's rendition of Fludd) and Galileo project their innermost possibilities of understanding on the elemental make-up of the world, assigning to the light carried by the expansion of *anima mundi* the preeminence of a constitutive first principle and possibly intimating its substantial co-extensiveness with God. I would not want to argue that theoretical commonalities or elements of doctrinal superposition are all important or even decisive. I think that the question of doctrinal belief is difficult to formulate, and that it depends on ontological commitments that are not easy to bring to light. It depends on what one thinks the being of mind and the being of belief really are at the fundamental level. Gassendi himself sensed that this was indeed the case as he admitted at the beginning that a thorough engagement with Fludd's system was necessary in order to penetrate and then criticize his mind. Engaging in reading Gassendi's dissertation with a view to understanding Fludd's principles opens a new vista on one's thought and on the thought of the 'other', and hence on one's deepest ontological commitments.

Did Galileo himself believe in *anima mundi*? The question requires thinking about the 'self'. It is not a thing like a stone. I prefer to speak of being-in-the-world as being prior to the self in the sense that it does not separate the self from the world. Galileo found his self wandering alongside sensations of pleasure as he read through Gassendi's dissertation. It is like the resonance between two vibrating strings, it is not so much absolute pitch that matters as sameness of pitch. Primordial tuning makes existence resonate rather than believe. Self and world manifest themselves in the tension of pleasure and pain.

Gusto

Something more should be said about Galileo's sensations of pleasure. His language is precise. He talks of the 'gusto' that reading Gassendi gave him. It is the gustative sensation that he wishes to emphasize as pleasurable. Hence it is gusto that has been mobilized by the reading. We feel the pleasure of gustative sensations in eating or drinking good foods and drinks. When we do so we replenish ourselves. The material that we ingest is altered by the processes of digestion, and by the physiological reactions of the body when it is exposed to intoxicants such as alcohol. Metabolism is the transformation that reading sets in motion when we ingest good words. Unlike other senses such as sight, which presupposes an object of sensation that becomes the basis for perception and knowledge, the gustative sensation need not become an epistemic process in the sense of an adequation of percept and the object of perception. We replenish ourselves by ingesting foods because we feel empty. We satisfy the need for food that triggers our hunger. Good food gives us pleasure. Gassendi's dissertation was such good food for Galileo. It did not need to become theory and knowledge. It needed to satisfy a need for good words that had been triggered by an unfathomable hunger.

Bodily Blindness, Light, and Geometry

Galileo became completely blind toward the end of his life, sadly before he could see a printed copy of his *Discorsi* (1638).[41] On 2nd January 1638, Galileo wrote his most beautiful letter on blindness to his friend, Elia Diodati, in Paris. He told Diodati that he had become completely blind about a month earlier. He continued with these words: "Now, Sir, think what a slough of despondency I find myself in, while I reflect that…that heaven, that world, that universe, which, with my marvelous observations and clear proofs, I had enlarged more than a thousand times beyond what had been seen by the sages of all times past, has now shrunk to the space occupied by my own body. The novelty of the incident has so far prevented me from accepting the fact, with the necessary patience. Assuefaction will only come with time. This extraordinary transmutation (*strabocchevole trasmutazione*) has caused in my mind a marvelous metamorphosis of thought, purposes, and concepts, about which I can only say little at the moment…".[42] In a few letters he exchanged with friends around that time, we find comments on his irremediably impaired capacity for deep thought, and especially on his compromised ability to pursue the search for geometrical proofs. Those comments stress the indispensable

41 Galileo had not yet received a copy of the book in March 1639 though it had long been available in Rome and elsewhere (Galilei 1890–1909, XVIII, p. 37).

42 Galilei 1890–1909, XVII, p. 247.

role that sight plays in connection with the visuo-spatial information afforded by geometrical diagrams. The "technology" of the illuminated diagram was essential in anchoring the technique of proportional reasoning, on which Galileo's mathematics was based, to visuo-spatial perception. In response to an inquiry by his pupil, Benedetto Castelli, Galileo wrote: "...and if I could regain a less troubled condition, I would explain to you my concept; but since it is a very complex excogitation, or structure, difficult to elucidate, especially with naked words, it being impossible for a blind person to draw a diagram, I am unable to say anything more specific, except that my strategy depend on a proposition by Euclid".[43] To Giovan Battista Baliani, who had sent him a new book on motion, Galileo, in 1639, replied: "...although I have been unable to understand the proofs clearly, since I could not correlate them with the diagrams".[44] To another correspondent, again in reference to Baliani's book, Galileo said: "...because I cannot see the diagrams and compare them with statement and proof, I remain doubtful in one or two places; since, I think, I failed to arrive with the imagination as far as the sense of sight would have allowed".[45] Perhaps the most extraordinary of Galileo's comments on cognition and blindness concerns his own proofs. Here is what he told Baliani in a subsequent letter. "Two other details which you mention in your letter I have been unable to compare with my writings, since there intervene linear diagrams and comparison of letters, which are impossible for me; and to my deep regret, I will forever be unable to understand even my own demonstrations, which contain diagrams and computations...".[46] When he succumbed to blindness, Galileo forever lost the ability to understand even his own proofs.

43 "...et io, se mai potessi ridurmi in stato men travaglioso, procurerei di significargli il mio concetto; ma perchè è una macchinazione o struttura assai grande e difficile a spiegarsi, e massime con nude parole senza poterne un cieco disegnare la figura, non posso per ora dir cosa essenziale, se non che il mio artifizio depende da una proposizione di Euclide" (Galilei 1890–1909, XVII, p. 360).

44 "...anchorchè io non abbia potuto intendere distintamente le dimostrazioni, non potendo incontrarle con le figure" (Galilei 1890–1909, XVIII, p. 11).

45 "...per non poter veder le figure nè riscontrarle con la dichiaratione e dimostratione, mi lascia in qualche scrupolo in un luogo o due; credo, per non haver potuto arrivare con la immaginativa sin dove il senso della vista vi si ricerca di necessità" (Galilei 1890–1909, XVIII, p. 37).

46 "Due altri particolari che ella tocca nella sua lettera, non ho potuto riscontrarli in quello che scrivo, intervenendovi figure lineari e rincontri di caratteri, impossibili essere da me fatti, come per mia infelicità resto privo di poter mai più intendere le mie medesime dimostrazioni, dove intervengono figure e calculi..." (Galilei 1890–1909, XVIII, p. 95).

I believe that Galileo's comments indirectly reveal a structural dimension in his life-world, the visuo-spatial mode of being, i.e., typically, the material presence of the diagram on an illuminated piece of paper. It speaks volumes about the impossibility of separating the world from the self. Geometry was a mode of being that presupposed light. There is no such thing as geometry in darkness. Remember that Galileo compared the human intellect's capacity for understanding geometrical proofs with God's non-temporal vision of the totality of geometry. Light is the vehicle that carries the human mind into the divine mind through the illuminated diagram. Blindness shrinks the infinite universe to the confinement of the bodily interior. Being-in-the-world is deeply affected by the *strabocchevole trasmutazione* caused by blindness. The loss of access to the illuminated diagram is not only an impairment for geometrical discovery but an existential transformation that modifies the whole self by clamping its authentic possibilities of understanding. The word *strabocchevole* means overflowing. The physiological transformation that precludes access to illuminated objects bursts the banks of the self. It inundates the planes of being submerging the ego. It returns existence to a primordial state preceding individuation. Light is the elemental matter that receives individuated geometrical forms. The blind person loses the possibility of understanding as mode of being in proximity with geometrical forms. This is what caused the total disorientation of which Galileo speaks with his friends.

· 5 ·

JOVE

Antonio Favaro was flabbergasted when he found among the Galileo manuscripts preserved in Florence a strange poem laboriously reworked.[1] Who was the author? The handwriting was Galileo's. But the verses had been sent him, apparently towards the end of 1610, by the Medici court poet Andrea Salvadori. The surprise was due to the fact that the autographs indicated that Galileo had elaborated Salvadori's verses in a highly original fashion. The inevitable conclusion drawn by Favaro was that Galileo had appropriated the poem, and that in consequence the authorship of the extant verses was to be attributed to both Galileo and Salvadori.[2] In fact, we do not even know whether Salvadori's original verses have survived. On the basis of corrections, additions, and deletions Favaro and Vaccalluzzo surmised that Galileo cobbled together a final version starting from Salvadori's originals.[3]

---

1 "Invece questa canzone si trova (chi avrebbe potuto immaginarlo?), scritta e riscritta di mano di Galileo…" (Galilei 1890–1909, IX, p. 234).

2 "Cosicchè questa poesia ben si può dire opera di due autori; del Salvadori, di cui crediamo sia l'intelaiatura e la prima versificazione; e di Galileo, che a sua posta la elaborò" (Galilei 1890–1909, IX, p. 235).

3 Galilei 1890–1909, IX, pp. 233–236, and Vaccalluzzo 1910, pp. 3–5. The poem was eventually published in a collection by Salvadori's son in 1668 (Vaccalluzzo 1910, p. 3).

Be that as it may, it is clear that the poem stirred up the motions of Galileo's soul. It consisted of twenty stanzas recounting the myth of the giants assaulting the throne of Jove. The poem celebrated their doomed feat before evoking, in the last part, a mysterious figure. The latter was to suffer the same fate for attempting to steal from Jove the four companions discovered by Galileo with the *perspicillum*.[4] The poem was a feat of *concettismo*, brilliant in its own right, but which must have scandalized Favaro. It has never been taken seriously as an object of historical-philosophical scrutiny. Propaganda, encomiastic rubbish, and all that jazz? Perhaps, to some extent. Yet the unbridled ego raging against the mysterious figure in the last three stanzas casts a shadow—Galileo's shadow.

The poem punctuated Galileo's exercise of silence, his withdrawal from the public arena, where, soon after the publication of *Sidereus nuncius* in 1610, the existence of the Medicean planets had been called into question dramatically. It seems as if a poetic madness takes hold of the ego. The archetypal world that preceded the emergence of humans breaks through the mask of reason, suspending the dialectic of argument and counter-argument. Before the insanity of the tongues jarring Jove, the ego appears to have lapsed into a state of mortification, dissolving into the unspeakable.[5]

Canzone per le stelle Meidcee temeriamente oppugnate	*Poem for the Medicean stars audaciously attacked*
Miseri, qual follia l'alme v'ingombra, Onde le lingue insane Ingiuriosi contra 'l ciel volgete?	Fools, what madness seized your souls, Whence the insane tongues you turn Offensive against the heavens?

To all intents and purposes the *perspicillum* worked like an alchemical retort. For the adept Galileo it prompted a regression to a modality of being-in-the-world where poetry is the means of responding to the primordial anxiety aroused by Jove's thunder—i.e., the cosmic rage overwhelming originary humans who were emerging from bestiality into consciousness. For his opponents, as we shall see, the iridescences visible through the lenses of the *perspicillum* were hallucinations to be explained by optical theory. In other words, the poem brings Galileo face to face with his shadow, the hubris of the non-human animal rising against the jovial hence discursive reason of humans. In *Sidereus*

4 Cf. Vaccalluzzo 1910, pp. 5–10, and Galilei 1890–1909, IX, pp. 267–272.

5 Vaccalluzzo 1910, p. 10 (first three verses of the penultimate stanza).

nuncius, the starry messenger described the splendor filling the air at Cosimo's birth in order that Jove's universal power, i.e., *clementia, mansuetudo, suavitas, maiestas, auctoritas*, the pantheon of pagan reason, together with a Christian soul already embellished by God with the noblest ornaments, could be seized by the delicate corpuscle of the newborn at the instant of his first breath.[6] Galileo, the giant armed with a magical *perspicillum*, who had violated the throne of Jove, now confronted his own shadow by abluting the desecration in a poem replete with *concettismo*.

Because of Christianity's doctrinal commitment to the ideal of the uniqueness of truth, as championed in the Augustinian controversialist inquiry *de vera religione*, and reaffirmed in the aftermath of the emergence of confessions in Europe, seventeenth-century reason turned into an institution of violence. Both Galileo and his opponents participated in that same institution that negated the plurality of theological and philosophical truths. The discovery of the Medicean *planets*, i.e., the fixation of the indeterminate character of the appearances as having the nature of planets, claimed by Galileo, was an act of surrender. It implied the handing over of the magical *perspicillum* to the institution of violence that had superseded the pluralism of the Graeco-Roman pantheon. Galileo's insistence on the exclusive definition of the Jovian observables in terms of *planets*, with calculable circular motions corresponding to those already known of the seven planets of tradition, determined a situation of break-down of the boundaries separating natural philosophical from theological argument. The separation had been worked out at Padua by the radical Aristotelians who opposed the Christian reading of Aristotle's doctrines on the soul. Galileo's *perspicillum* brought down a cultural barrier slowly erected at the University of Padua in the course of three centuries of exegesis of the Latin corpus of Aristotle's works. Unbeknownst to Galileo and his contemporaries, the *perspicillum* (which, unlike modern psychotropic substances that have been demonized in most democracies, was never made illegal by the Roman Catholic Church that condemned Galileo) became the technological symbol of an unprecedented form of institutional violence which, in both Western science and Christianity, has since silenced the pluralism of veridical voices, that is, the plurality of gods speaking their conflicting truths, and thus weaving the fabric of mythological actions that challenge dogmas. I have sketched my pathway towards Galileo's shadow prospectively. With some liberties I follow in the footsteps of Giambattista Vico's theory of poetic

6 Galilei 1910, pp. 3 recto–3 verso.

characters.[7] Vico's intuitions in *La scienza nuova* (The new science) will escort my groping, here and there, in the chapter.

Hallucinations, Observables, and Imaginative Universals

Pagan societal institutions, says Vico, were founded when the early giants began to turn into humans, loosing their disproportioned stature, by dint of sacrifice, ablutions, cleanliness of body, and the maturation of a sense of reverence for the gods. As Donald Verene pointed out, in Vico's view, "there can be no institutions or society of giants".[8] The creature armed with the magical *perspicillum* regressed to a mythical pre-human state among giants. By "fleeing from the wild beasts with which the great forest must have abounded, and by pursuing women, who in that state must have been wild, indocile and shy, [they] became separated from each other in their search for food and water. Mothers abandoned their children, who in time must have come to grow up without ever hearing a human voice, much less learning any human custom, and thus descended to a state truly bestial and savage…They would be quite without that fear of gods, fathers and teachers which chills and benumbs even the most exuberant in childhood".[9] That disproportionate and wild Galileo, slumbering in the unconscious without fear of gods, turned into the *human-all-too-human* messenger with an act of renunciation of the plurality of veridical interpretations of the Jovian observables. That sacrifice inaugurated the novel institution of telescopic astronomy by devolving the *perspicillum* to the monism of the Christian institutions of knowledge.

---

7 I am inspired especially by Donald Verene's interpretation of Vico's poetic characters as imaginative universals (Verene 1981, pp. 65 ff.).

8 Verene 1981, p. 93.

9 Vico 1948, p. 101. In his powerful vernacular Vico's original passage reads as follows. "… per campar dalle fiere, delle quali la gran selva ben doveva abbondare, e per inseguire le donne, ch'in tale stato dovevan esser selvagge, ritrose, e schive; e sì, sbandati per truovare pascolo, ed aqua, le madri abbandonando i loro figliuoli, questi dovettero tratto tratto crescere senza udir voce umana, nonchè apprender' uman costume: onde andarono in uno stato affatto bestiale, e ferino…e senza alcuno timore di dei, di padri, di maestri, il qual' assidera il più rigoglioso dell' età fanciullesca" (Vico 1744, I, pp. 133–134). Note that I have avoided repeating the numerous italicized phrases and capitals in the original text for the sake of readability.

Yet the indeterminacy of the observables, originating in the regressive bestiality that was exalted in the poetic imagination by Galileo-Salvadori, had emerged into the market place, the Vichian great forest of modernity, immediately after the publication of *Sidereus nuncius*. Without fear of fathers and teachers, the beast spoke publicly through the voice of a young student who had for some time been in the private service of the Bologna astrologer Antonio Magini. It was the voice of the Bohemian Martinus Horky, who roughed up minds when he published a satirical *Brevissima peregrinatio contra nuncium sidereum*.[10] In anachronistic philosophical parlance, one might say that its focus was the epistemic status of the Jovian observables. But it took the form of a narration of a journey, as the title announces. Its substance, bubbling in the *nigredo* of the mind in which Horky's experiment was caught, was the inarticulate noises of the regressive human animal seeking a first thought, a movement towards individuation, as a response to the reawakening of a primordial fear of the gods. Verene comments that "Vico's view of fear suggests something broader than just a theory of the first thought. It implies that fear is the condition, the impetus for any first thought. Any beginning in thought subsequent to the primordial thought of the first men reenacts this fear of the beginning. The intelligible universal of modern men is not connected with fear. It is thought confident of itself. In Cartesian fashion it moves from one concept to another, but it does not make its own beginnings".[11] It is precisely this moment of a reinitiation of thought, when thought is no longer confident of itself, because it has shed the mask of violence that legitimates the logic of the movement from concept to concept, that makes Horky's satirical experiment with the Jovian observables so intriguing.

The stillness of thought, the paralysis of the minds stunned by the resurgence of anxiety that followed in the wake of the *Sidereus nuncius*, was vividly

---

10 The rare *Peregrinatio* was republished in Galilei 1890–1909, III. 1, pp. 127–145. Horky was apostrophized by Favaro as "…boemo detrattore di Galileo, altrettanto burbanzoso, quanto cattivo scienziato e latinista pessimo" (Galilei 1890–1909, III.1, p. 11). To Favaro's credit it must be said that he had been the first to include Horky's *Brevissima peregrinatio* in an edition of Galileo's works. See Westman 2011, pp. 468 ff., for a tabloid chronicle that entangles the vicissitudes of Horky with the myth of the scientific revolution ("… Copernicus's formidable position in the history of astronomy and in the historiography of the Scientific Revolution is hardly open to dispute", *ibid.*, p. 28), and Strebel 2006, for a summary exposition of the arguments put forward by Horky. Virtually nothing in the way of sustained historical-philosophical inquiry can be found in the secondary literature concerning Horky's response to Galileo's surrender of the Jovian observables.

11 Verene 1981, p. 93.

described by John Wodderburn, who published a refutation of Horky's *Peregrinatio*.[12] In the preface to the reader, Wodderburn emphasized that the *Nuncius* had appeared to shake the foundations of astrology, and above all that it had plunged the minds into a stupor comparable to that which the early humans must have felt when they observed an eclipse of the sun and the moon emerge from the obscurity of night.[13] Since, as Wodderburn pointed out, no mathematician had plucked up the courage to speak publicly, for or against Galileo's celestial novelties, he decided to break the silence in defense of Galileo.[14] He cautioned, however, that Horky's *Peregrinatio* was written not so much as a philosophical work but rather in the style of a satirical poet.[15] What occasioned Horky's *Peregrinatio*?

On April, 25th and 26th, 1610, Galileo demonstrated the *perspicillum* in Bologna to local notables. Horky had the opportunity to examine the instrument and make use of it at leisure during two nights. He reported that the first night he saw two spots (*maculae*) close to Jupiter, and that when he asked Galileo where the other two were supposed to be, no answer came, after which they departed in silence.[16] The second night he was able to observe the four *maculae* successfully.[17] He claimed that the Jovian observables, the *maculae*, were the result of an hallucination due to some sort of concentration of the celestial light visible through the *perspicillum*.[18] Horky engaged in a phenomenological description of his stupefying experience, reporting that he decided to redirect the *perspicicllum* slowly away from the body of Jupiter towards the position where one of the *maculae* had appeared. He stated that as soon as he aimed his gaze at the spot where the *macula* had been visible, it was Jupiter that manifested itself, precisely there, while the *maculae* had evanesced. He styled his observation of the Jovian observables a *true ocular demonstration*.[19]

---

12 On Wooderburn, cf. Spini 1996, pp. 79–89.

13 "…et quatuor novos introduxit planetas, astrologiae peritiam et principia prima ex instituto subvertere visus est. Nec minus omnes stupore afficiebantur, quam primi illi hominum, qui solem eclipsari et lunam nocte involvi conspexerunt, tanti naturae miraculi causam ignorantes" (Galilei 1890–1909, III. 1, p. 153).

14 Galilei 1890–1909, III. 1, p. 153.

15 "…in reliquo tempus terit in rixis et iocis, ut non tam philosophari, quam poetae satirici partes agere, videatur" (Galilei 1890–1909, III. 1, p. 162).

16 Galilei 1890–1909, III. 1, p. 140.

17 Galilei 1890–1909, III. 1, p. 141.

18 Galilei 1890–1909, III. 1, p. 141.

19 "Sed, cum perspicillum lente a corpore Jovis amovens, centrum hoc, ubi novos planetas videbam, quaesivi, tum statim mihi surrexit Iupiter, et non erat his novus planeta; unicus

In view of the well-known neurophysiologic fact that human visual acuity is located in a restricted central area of the retina, and that the human capacity for integrating visual scenes is a fantastic construct of our mode of being-in-the-world, remarkably reliable under favorable circumstances but nevertheless relying on scant sensory information, Horky's phenomenological description of the vanishing *maculae* from the field of vision has to be taken very seriously. Remember that he had not been exposed to the Jovian observables before the Bologna demonstration. Perception is a learned modality of being-in-the-world that is open to interpretation. Further, Horky claimed that he could see a candle flame doubled through the objective lens of Galileo's *perspicillum*, and that he and others had seen *Spica Virginis* (the brightest star in Virgo, also known as *Alpha Virginis*) duplicated, like two stars, through the *perspicillum*.[20]

Horky's ocular demonstration of the indeterminacy of the Jovian observables manifests thought that has lost confidence in itself, that has turned young again, hence thought that is making its own beginnings anew, groping towards becoming logical thought. It manifests thought caught in its forward-thrusting movement towards individuation. But the atmosphere of the vessel in which the Jovian observables had been thrown remained caliginous. For Horky's satire still disqualified his visions by appealing to hallucination, and desperately invoked the violence of reason's coming to the rescue of his ocular demonstration, namely, reason in its Christian apotheosis, theological, philosophical, mathematical, and optical (the hierarchical ordering of the sciences alluded to by Horky is revealing of his hapless state of mind).[21] Horky was able, albeit only briefly, to live through the imaginative universals of the human collective unconscious, i.e., the emotional experience of the forces of nature, such as Jove, that gives rise to intelligibility prior to the irruption of concepts.[22] He stared at the empirically given in its brutal face. He was stupefied but docilely coiled away from the abyss that opens up the possibility of human individuation.

That is why I intimated at the beginning that his satirical experiment was the stage of a *nigredo*. The furnace cooled down too quickly, and Horky's

---

enim Iupiter restabat, reliqui autem quatuor famuli Ioviales ultra polos avolarunt. Hanc meam opinionem et veram contra novos hos Planetas ocularem demonstrationem…" (Galilei 1890–1909, III. 1, p. 141).

20 Galilei 1890–1909, III. 1, p. 142.

21 Galilei 1890–1909, III. 1, p. 142.

22 Verene 1981, p. 71.

encounter with the imaginative universal of Jove concluded with an ambitious metaphor, worthy of a *concettista*, borrowing from Ovid's pagan conception of the creation of man. He insisted that the Jovian observables, the four *maculae* obscuring the field of vision, likely caused by a combination of effects due to refraction of light and atmospheric caliginousness, must captivate the mathematicians who gape at heavenly marvels. For, human beings have been granted by the creator a face to gaze upon the heavens, their future abode, unlike non-human animals, irrevocably face-less, that must forever turn their heads downwards to scrutinize the surface of the earth.[23] The human gaze pierces the heavens through the *perspicillum* to admire the celestial miracles encircling Jove, while here on earth miracles illuminate divine intellects between which only posterity will be able to adjudicate. We are all divine intellects, Horky seems to be saying, insofar as our gaze turns to the heavens, ecstatic, but we are all prone to retrograde, like the planets do, into earth-bound animality when we turn our gaze downwards. Humanity is not a stable modality of being-in-the-world. It is an unstable process of circulation periodically subject to retrogression into non-human animality. The non-human animal is destitute of the degree of freedom of verticality. It is dimensionally poor in the world. Its mode of presence in the world is horizontal. The human is a living emblem of the cross, hanging downwards towards perdition but with the gaze upwards and the arms wide-open to embrace fate. There is no this-worldly hostility in Horky's final appeal to posterity.

The mechanistic-materialistic physiology of the senses that was established in the nineteenth century in the wake of positivism, and which still defines contemporary neuroscience, has lead the modern imagination to a misconception of the organs of sense, fancying perception in terms of a process by which organisms respond to stimuli in the environment. Perhaps this explains our neglect of Horky's satirical response to Galileo's recantation of the pluralism of visions.

This view of perception labors under the ontological presupposition, enforced by positivist science, that stimuli and perception are processes existing in the world independently of each other. Stimuli are supposed to be caused by physical objects, which are external to bodily boundaries defining the perceiver, while perception is to be explained by internal mechanisms that have been fashioned phylogenetically by adaptation and struggle for survival. But, if perception is an originary mode of being-in-the-world, prior to the

23 Galilei 1890–1909, III. 1, p. 145.

separation of human from non-human, in which imaginative universals, the aesthetic manifestations of the primordial forces of nature, freeze the current of sensations, as I think it is, concurring with Verene's reading of Vico, then the emergence of the Jovian observables in the field of vision must be seen as a creative modification within the one being, the one world of the heroic pantheists, such as Giordano Bruno, or the unus-mundus of Carl G. Jung. In this perspective, there is room for allowing the satirical dissociations of Horky's journey with the sidereal messenger to be true—as he claimed his *demonstratio ocularis* to be a *vera demonstratio*—in the sense of truth as indivisible whole appearing to us in the guise of the infinite modifications of being that give rise to the illusion of parts. There is room, in other words, for preempting the raping of the world-soul that the institutions of knowledge-as-violence, patronized by the dialectical transformation of Catholic and Protestant fundamentalisms into the objective spirit of the mechanistic world-view, have continually perpetrated.

Pathologies of Vision

Why had Horky seen a candle's flame duplicated through Galileo's *perspicillum*, a phenomenon that cast doubt on Galileo's interpretation of the Jovian observables as planets? Wodderburn answered this disturbing question with an attempt to pathologize the phenomenology of vision. Nobody had claimed that the instrument duplicated opaque objects. But with an object possessing light (*obiectus luce praeditus*) he was convinced that deception is possible. Such an object can appear double when the perpendiculars to the planes of the lenses of the *perspicillum* do not coincide. In this case, he argued, the species reach the eye twice, that is, both through the plane and the concave surface of the eyepiece. He added that to avoid this error some used to make the plane surface of the eyepiece impenetrable to vision like white ice.[24] So far Wodderburn has pointed out one possibility of deception due to inaccurate alignment of the radii of the spherical segments on which the two lenses have

24 Galilei 1890–1909, III. 1, p. 169. I am not sure I understand what procedure Wodderburn describes here. I think the whitening process he suggests could be done by moderately grinding the plane surface of the eyepiece lens. If this is right then it is possible to speculate that double vision in early *perspicilla* was a rather common phenomenon, and that techniques had been adopted to correct the error. The question remains as to why Horky saw a double candle with Galileo's own instrument, though perhaps he did not reassemble it correctly after examining its lenses.

been grinded. But then he asserts that Horky could not have seen a candle duplicated in any other way.[25] Why not? This intolerant conclusion paves the way for epistemic repression to take the form of an imaginary pathology of vision. If the *perspicillum* is correctly assembled and if you still see a candle's flame double, then you must be visually impaired.[26]

But perhaps the issue is not healthy vision, it is the category 'deception'. Why does Wodderburn insists on deception? Why is duplicity of appearances intolerable to Wodderburn and to some of us as well? Horky solicited our gazing at the heavens to encounter the celestial admirables (*miracula*). He had gazed at Jovian admirables rather than look at the Jovian planets. The motion upwards of the gaze separates the human from the non-human animal turning its head downwards. The Jovian admirables mark off the momentary humanity of Horky from the non-human animality of Wodderburn. Galileo's defender turns his head downwards writhing his gaze away from the celestial admirables. He refuses to gaze in exchange for earthy looking. The early modern enforcement of looking became foundational for modern science. It relegated the gaze to a pathology of vision from which we have not yet recovered. The promotion of looking entails the elimination of the pluralism of visions. It is the precondition for the constriction of theory into an incantatory logic that imputes an abnormality to the victim of violence. Thus the victim that Wodderburn put under scrutiny was the gaze. He wrote a mourning-piece featuring the mystery of light—philosophers of science would call it a theory of light—as a memorial of the dying gaze. I find it unnerving that at its place of origination modern scientific theorizing is a mourning that, however, if recovered from oblivion, might turn into a therapeutics for our dying collective self-knowledge. (At least it has been so for me as I began to learn, while researching this book, how to turn my gaze towards my own shadow).

---

25 "...neque alio modo poterit nobis ostendere peregrinus noster quomodo candela accensa appareat geminata" (Galilei 1890–1909, III. 1, p. 169).

26 This is Wodderburn's conclusion, as he arrogantly addresses Horky this way. "Sed forte obtusam habes aciem?" (Galilei 1890–1909, III. 1, p. 172).

And He Made a Mourning

For Wodderburn the error that Horky made was one of fundamental principles. It was the assumption that luminous bodies that are subjects of light such as the sun give off light in the same way as fire, which emits corpuscles in the form of sparks. This principle, says Wodderburn, has long been exploded as paradoxical. Luminous bodies do not illuminate by diffusing an effluvium of bodies, for light is drawn out from the potentiality of matter, which is a diaphanous body, exactly like heat is drawn out from the passive body, with the important difference that heat, unlike light, does not emerge wholly in a single instant.[27]

This theory of light based on Aristotle's conceptual framework rules out the possibility of drawing a valid analogy between the light of celestial bodies and the light of a candle's flame. The possible duplication of the latter as seen by Horky through the lenses of the *perspicillum* does not imply that the *perspicillum* will duplicate the images of celestial bodies since these give off light of a totally different nature. But Wodderburn does make concessions. He concedes that, as Horky suggests, light migh be somehow concentrated in the vicinity of a luminous body such as Jupiter, from which effect Horky would like to conclude that the light of Jupiter might be diffracted by the lenses so as to split its image or even multiply it. But since, for Wodderburn, light is unlike the material effluvia given off by fire, but rather a quality of the diaphanous body educed from it instantly, it cannot be concentrated in the vicinity of Jupiter in accord with the *uniformly difform* action of qualities that increase temporally, i.e., successively, part after part, uniformly.[28] In Wodderburn's medieval natural philosophy of qualities intensifying *uniformly difformly* (uniformly accelerating, will say Galileo in reference to the quality of speed), there is no explanation in terms of temporal action for the instantaneous process of intensification of light educed from a diaphanous body. Hence Wodderburn has to concede that other factors might explain the possible intensification of celestial light in close proximity of the celestial bodies. These include the great distance of the luminous body from us, any indisposition of the medium, excessive intensity of light, as is the case with the sun, or impaired vision. All these factors can be subsumed under the general principle that the luminous object becomes disproportioned with respect to the visual faculty.[29] If

27 Galilei 1890–1909, III. 1, p. 170.

28 Galilei 1890–1909, III.1, p. 170.

29 Galilei 1890–1909, III.1, p. 170.

this disproportion between the organ of sense and the object of perception is suspected then appearances might be anomalous.

Wodderburn has come full circle, I think. He has ruled out the analogy flame-Jupiter elegantly. But he has conceded that no explanation of the intensification of light is available in his conceptual framework. Further, he has acknowledged the problems of distortion that might be caused by a variety of intervening factors. Finally, he has been left with no alternative but to resort to the proportionality of the subject and object of perception as the ultimate court of appeal for judgments concerning the veridicality of ocular observables. I think he has de facto conceded defeat. He must defend the veridicality of the Jovian observables by assuming without proof the correct proportionality between the observer and the observables. There is no provision in his conceptual framework for demonstrating the correct proportionality of the subject and object of perception in cases of vision, such as observation with the *perspicillum*, that exceed customary experience. Only protracted experience with *perspicilla* will tell us whether that proportionality can be reliably established on empirical grounds.

I hinted above that theory-building is more like a mourning, and that we can recover its meaning from oblivion for therapeutic purposes. Why do we build natural-philosophical theories in the first place, such as the late-medieval theory of *uniformly difform* qualities invoked by Wodderburn? Because, I suggest, theorizing as a mourning defuses our anxiety about the death of the phenomenon, a death occurring both in Christian-Aristotelian natural philosophy (brought about by the renegation of the world-soul as the receptacle of phenomenal qualities that was taught by Hermeticism) and in modern scientific inquiry (where it was brought about by the advent of the mechanistic worldview). Thus, Wodderburn mourns the death of the gaze, the exceptionality of humans admiring the spectacle of the phenomenal world. He does so paradoxically by theorizing about light.

The mechanistic world-view inured us to insensibility towards the phenomenal, and I believe that it was not by accident that early twentieth-century quantum physics exploded that world-view by reinstating a notion of spirit, or world-soul, in the guise of the more secular notions of field and quantum wholeness. At the beginning of the twenty-first century, our desensitized souls are in need of recovering from profound oblivion the pleasures of the gaze, revivifying the magic of the imaginative universals that are the perennial rejuvenation of thought.[30]

The Sacrifice of Friendship

On April 16th, 1610, Horky wrote a letter from Bologna to Johannes Kepler asking him if he had been able to see the Jovian planets.[31] When Galileo left Bologna on 26th April, Horky wrote again to Kepler in a satirical mood, reporting that he and others had seen fixed stars doubled with Galileo's *perspicillum*, and that eventually Galileo had departed in a slough of despondency, following his failed demonstration, and without even thanking his host Magini for the many ideas he had received in exchange for peddling a fable.[32] On May 24th, Horky told Kepler that he had written harshly against *Sidereus nuncius* but that since Galileo had many friends he had decided to change tactic and rewrite his piece in a style similar to Kepler's *Dissertatio*.[33] Finally, two days later, Horky announced to Kepler he was going to send him a draft of the *Peregrinatio* for review. He made good on his promise on June, 30th, with a letter from Bologna.[34] In the meantime, Kepler warned him that he suspected that Galileo knew well that the *perspicillum* might double images, and that the reason why people in Bologna were deceived was that while Galileo had perfect eyesight they were all myopic. Kepler even suggested that to obviate the problem of double vision the eyepiece should be adapted to each observer so as to correct his defect.[35] Another instance of pathology rushing to the rescue?

30 May I hint, with apologies to the reader, at my juvenile experiences with my first amateur telescope in years gone by, when I marveled at the vision of the Jovian four points of light as endless possibilities of thought, a mystery not to be dissolved by the consciousness of their moon-like nature.

31 Galilei 1890–1909, X, p. 316.

32 Galilei 1890–1909, X, pp. 343, 386.

33 "Scripsi enim durissime contra Nuncium Sidereum: illa omnia Nuncii huius pater, me inscio, cum in nostra domo Bononiae pernoctatus est, abstulit. Quia autem multos amicos hîc habet, muto animum, et, secundum Dissertationem tuam doctissimam, forma aliam sequar " (Galilei 1890–1909, X, p. 358). I am at loss as to what Horky meant by stating that Galileo had snatched away "illa omnia" overnight, unbeknownst to him, during his stay at Maginis's house. Might Galileo have stolen Horky's manuscripts? It seem preposterous yet not impossible given the climate of hostility and reciprocal diffidence in which the Bologna episode took place.

34 Galilei 1890–1909, X, p. 359.

35 Kepler 1937–2009, XVI, p. 315.

This time a social myopia seems to have effected everybody in Bologna except Galileo.

On June, 22nd, Giovanni Roffeni wrote Galileo that Horky had been in Modena to have the *Peregrinatio* printed there (as it eventually happened), and that when Horky came back to Bologna, he had threatened him, at Magini's, bidding him to withdraw what he colorfully called a 'maledica scrittura' (vicious piece of writing). Moreover, Roffeni reported that Magini had kicked Horky out of his house, and that on the same evening Horky had been spotted, distraught and in a state of psychological distress (*....tutto malandato e disperato*), on his way to Modena.[36] August, 9th, 1610, must have been excruciating for Kepler. On that day he wrote to both Horky and Galileo. To the latter he said that he had received from Horky, whom he qualified as incompetent (*imperitus*), the *Peregrinatio*, on whose pages, he added, one would waste time.[37] Kepler remarked emphatically that Horky had assassinated their friendship at the outset in a most slanderous manner.[38] He then beseeched Galileo to name the witnesses to his observations that he had mentioned in correspondence.[39] To Horky Kepler wrote bluntly that the fame of his candor had become incompatible with their friendship.[40] On August, 19th, Galileo answered Kepler with a contemptuous letter in which, after deriding Horky, he ridiculed Kepler's request for witnesses with a laugh of mockery. What was Kepler talking about? Galileo was now the philosopher to the Grand Duke of Tuscany, with a princely salary, and with no obligations other than write his books on mechanics, cosmology, and physics. Was Kepler asking for witnesses? "Me ipsum!", exclaimed Galileo, myself!, who have been offered a salary by the University of Padua that no mathematician could have dreamed of. But the most affronting claim of the letter was yet to come. Galileo gave Kepler to understand that his tenure at Padua was secure regardless of whether the planets turned out to be fleeting illusions.[41] He then

36 Galilei 1890–1909, X, pp. 375–376. The episode was confirmed by Magini, who went so far as to suggest that in retaliation Horky should be beaten with a cudgel, "e la risposta a costui sarebbe di farlo bastonare…" (Galilei 1890–1909, X, p. 377).

37 Kepler 1937–2009, XVI, p. 320.

38 "Amicitiam hanc, inquam, vix dum spirare visam obscurissime, nece famosissima jugulavit" (Kepler 1937–2009, XVI, p. 320).

39 Kepler 1937–2009, XVI, p. 322.

40 Kepler 1937–2009, XVI, p. 323.

41 "Me ipsum produco, qui, in hoc Gimnasio stipendio insigni florenorum 1000 decoratus, et quale mathematicarum professor nullus habuit unquam, et quo tuto, dum viverem, frui possem, etiam illudentibus planetis et effugientibus…" (Galilei 1890–1909, X, p. 422).

mentioned that, apart from Giuliano de Medici, many others had seen the Jovian planets but did not dare to speak out. Finally, on October, 25th, 1610, Kepler communicated with Galileo about his meeting in Prague with Horky, who had paid him a visit on his way back home from Italy. The spectacle of officiousness was bizarre, said Kepler, who portrayed Horky approaching him, triumphant in his face, as though he had defeated Galileo, and speaking to him in a tone of great expectancy about his approval. Then Kepler answered to Horky that he had renounced his friendship with him. It was a moment of revelation that, Kepler remarked, had disturbed both of them, Horky being taken aback since Kepler's letter, in which he had told Horky about his decision, had reached Bologna after the latter's departure. He reported that when Horky expounded his arguments to persuade him to change his views, he crushed them rather than respond to them.[42] But Kepler also pointed out that Horky was genuinely convinced of the consensus against the Jovian planets that he had gathered from academics in Italy, and of which he produced evident documents.[43]

The connection between friendship, persuasion, and physical or verbal violence in early modern scientific argumentation has never been a subject of historical-philosophical investigation. Horky was abused twice, once in Bologna, perhaps even physically, at Magini's house, and again in Prague, verbally, by Kepler's assault on his arguments. In this regard, it is possible to make the suggestion that the discrimination between physical and verbal abuse is also doubtful in a context where the person attacked is especially vulnerable. Remember that Horky was very young and found himself among foreigners in Bologna. He was entrapped in a power-relationship between pupil and master while serving as an apprentice astrologer at Magini's house. The violent subjugation of the pupil by repeated verbal and physical aggression is in sharp contrast to the reemergence of the Jovian archetype constellated by the *perspicillum*. The young is seeking philosophical speculation, knowledge of self towards individuation, in the glaring light of the Jovian observables that have

42 "ego illa argumenta sui solvi, seu potius oppressi…" (Galilei 1890–1909, X, p. 457). Aviva Rothman has commented on the function of Kepler's rhetoric in his correspondence with Galileo, showing the fusion of argumentative and persuasive strategies in defense of Copernicanism (Rothman 2009). Perhaps the narrative of Kepler's verbal colluctation with Horky, the emphasis on the crushing of arguments rather than their solution, suggests a physical, bodily side of rhetoric that it would be rewarding to explore further.

43 "de quorum consensu fidem mihi fecit documentis manifestissimis" (Galilei 1890–1909, X, p. 458). It is hard to say what these documents might have been, possibly letters, or private notes.

come to share with him the space of the imagination. The aggression of phenomenal being's manifesting itself in the visions of the *perspicillum*, by dint of logical argument, destroys both the candor of Kepler's fame and the candor of the pupil in search of illumination.

Seneca teaches that revenge is an inhuman word, but the uttering of words that speak the renunciation of friendship on grounds of intellectual disagreement is worse than acting out the word 'revenge', it forces the human animal to renounce language's civilizing power to quench the thirst for revenge. It does not plunge the human into a brute, for that already implies the disqualification of the non-human as a carrier of the violence that humans themselves perpetrate on other humans and on other living beings. It is a pact with the devil's logic of argument and counter-argument that urges early modern thought to take revenge on dissidents. In observance of the dogma of the immaculate uniqueness of truth, promulgated by the emergence of the Christian denominations in Europe, the early modern ego was always ready to sacrifice friendship for the pleasure of winning the argument. Christianity bequeathed to modern science the supremacy of dogma, and with that came the disavowal of dialogue as a modality of being-in-the-world open to the differentiation of truths that places the interlocutors on the level ground of equality. Why did Kepler seek closure of the debate by renouncing friendship? Why did violence take hold of his ego, turning angelic will-to-reason into devilish will-to-power?

In early modernity the animal body, alive or dead, became a locus of violence in a bewildering variety of rituals. There was the dissection of corpses before a public and the vivisection of non-human animals, not to mention the fantastic procedures for disintegrating the body under torture in both public squares and secluded spaces, and then came the private rituals of human bodily self-mutilation, or self-flagellations, the mortification of the body as a discipline towards sanctity. The soul also became a locus of violence though the rituals followed different rules such as examination of conscience, spiritual exercises, the mortification of the intellect, the confession of sins, and especially the oath of allegiance.[44] Early modern science was hospitable to

44 The sixteenth and early seventeenth centuries saw the accomplishment of a transformation in the practices of allegiance, especially in early modern jurisprudences, legal structures, and strategies for disciplining the individual in relation to a collective body. Paolo Prodi has devoted a ground-breaking study to the transformation of the promissory oath, typical of the 'società giurata' of the late Middle Ages, into the ideologically-laden oath of allegiance (to a religious confession, a political party, an educational institution,

verbal and physical violence in a variety of formats. The sacrifice of friendship explored in this section was one. In examining its anthropology I have experienced a detoxification of existence, the beginning of a rehabilitation, while hoping to locate regions in our cultural landscape that welcome the presence of alterity without violence.

Philosophy as rehabilitation cannot unfold deductively since argument itself is the carrier of violence. Rather it should peacefully suffer the other emerging into experience and ideally allow multiple orders of perception to suspend the spatio-temporal order. Rehabilitation undermines the pursuit of truth as assimilation of the other by confronting the coercion carried by argument. It seeks catharsis by confronting the violence, both physical and epistemic, towards human and non-human animals, which has defined Western reason throughout modernity. Thus I renounce truth as assimilation of the other. I strive to resist the urge to spatio-temporal individuation. I make an effort not to privilege models of reasoning abstracted from the natural sciences. Rather I explore the orders of perception that constitute animal existence, endeavoring to enjoy the otherness of truth as is, diffracted through my own *perspicillum*.

Galileo's Shadow

The pleasure of winning the argument for its own sake is a disorder of the tongue, a pathology of speech that damages the soul. Hence Christian spirituality insists on the mortification of the appetite for argumentation, for disputatious verbal contest, namely, *l'appetito di contendere, et di contrastare troppo in parole*.[45] In this section, I wish to explore a hermeneutic of paradox, by matching words against Galileo's strange silence, his self-imposed mortification of the tongue in response to the adversarial speech about the Jovian planets that

---

etc.), with all its baggage of prescriptions and prohibitions to which the individual had to submit. Cf. Prodi 1992, pp. 264–269, where he discusses the approval by the Council of Trent in 1563 of an annual oath for all members of Catholic universities, an oath of allegiance to "ea, quae catholicae fidei sunt", i.e., a *professio fidei* comprising all truths that must be embraced and all heresies that must be rejected by Catholics.

45 Here I rely on the vast literature on spiritual discipline that was promoted by Reformed Catholicism, but particularly on the manual for achieving spiritual progress by the Spanish Jesuit Francisco Arias, which was popularized in early seventeenth-century Italy by Giulio Zanchini's translation (Arias 1602, II, p. 251, for the quotation).

followed in the aftermath of *Sidereus nuncius*. Galileo never answered Horky's strictures.

The hermeneutic that I am seeking here—by way of paradox, I reiterate—opposes the academic habit of "hiding one's own heart under the cloak of so-called scientific understanding [...], which fears and rejects with horror any sign of living sympathy [...] partly because an understanding that reaches the feelings might allow contact with the foreign spirit to become a serious experience".[46] This seriousness of silence that has become alien to understanding, in the face of the ridicule to which one attempting to revive its counter-values must expose oneself, is in sharp contrast to the decadence of an academic culture that has elevated argument for the sake of argument to the rank of a fetish, to an instrumentality in the service of career-managing in the market economy, that is, prostitution of educational principles for mercenary self-promotion. This seriousness can be recovered by observing one's thought processes in the act of puzzling over a sixteenth-century emblem of silence that catapults us into a Hermetic state of mind in tension (or harmony?) with the spirituality emphasized by Reformed Catholicism (fig. 5.1).

The problem at stake is truth, namely, the truth of the claims concerning the existence of the Jovian planets. Is truth not the outcome of argument? Even when argument is aimed at clarifying truth, through a legitimate process of learning, once the truth has been declared, the disputation has to stop whenever further reasons or authorities are lacking, and silence must prevail, for otherwise the contest will not only fail to manifest truth, but it will confuse and obscure it. Argument and altercation impede the act of reason, and therefore they impede the process of bringing truth to light, thus negating the value of reasoning.[47] It is human speech that must be mortified. For, the tongue, the organ of phonation, is worse than a wild beast, and while humans can tame brutes the tongue cannot be subdued by their powers, unless divine

46 Wilhelm & Jung 1931, p. 77. The insight is by Carl Jung. I partly owe my perplexity towards argument for the sake of argument to my former colleague, John Earman, a philosopher of physics and an inspiringly silent man who warned me long ago to beware of the 'superficial cleverness', as he put it, that is all too often on display during the question and answer sessions following talks, in which we, uneducated academics, show an amusing tendency to contrarianism that only manifests our blindness to psychological complexes larger than life.

47 "...perché la contesa non pure non manifesta la verità, anzi la oscura et la confonde. [...] La contesa, e 'l contrasto impedisce l'atto della ragione, et per conseguente impedisce il metter fuori in luce la verità, et toglie il frutto della disputa" (Arias 1602, II, p. 253).

Fig. 5.1. A representation of Hermes Trismegistus invoking silence as a means of worshipping, "Worship God in silence" (silentio Deum cole), from *Symbolicarum quaestionum, de vniuerso genere, quas serio ludebat, libri quinque* (Bocchi 1574, p. 138). The seven golden candlesticks might be a reference to the *Apocalypse of John*, unless it is the Hebrew Menorah. Above Hermes's head there is a Latin inscription, circularly written around a point at the center, *monas manet in se* (the one remains within itself), which is an allusion to the Hermetic monistic philosophical outlook. The Greek inscription at the bottom is a fragment by Simonides, Λαλήσας πολλάκις μετανόησε, σιωπήσας δε ουδέποτε: *It never repented him that he had held his tongue, but often that he had spoken* (English translation from Plutarch 1704, p. 246). The second poem accompanying the image, entitled *He whom divine mind will touch, must withdraw his mind from the senses*, refers explicitly to Hermes Trismegistus. "He will easily excel at all things who can withdraw his mind from the senses, and turn away his thought from habit, for he lives as mind, and as God. But he lives as a brute animal who lives by the body and the senses. He who remembers this judgment by Hermes Trismegistus will not be wretched". Cf. Bocchi 1574, p. 139. From: The *Library of the University of Illinois*, in the public domain at Internet Archive (https://archive.org/.).

intervention will help them.[48] Mortification is not only an exercise for the exploration of one's thought, geared towards the pursuit of truth in silence, it is a form of analysis of one's consciousness that must be brought to bear on one's whole personality. Thus, the senses must be mortified. They are the doors of perception, through which the vanity of our imaginations stirs up disorderly desires.[49] Curiosity itself has to be mortified, for it is a disorder of the sense of sight, as when, for instance, we are attracted by monstrosity for the pleasure of seeing what it is, or just seeing novelties.[50] The spectacles of nature are permissible, indeed profitable, such as the order and beauty of the universe, the light of the sun, the globe of the moon, with its variety of shapes, the alternation of day and night, and the courses of the stars, as well as terrestrial spectacles, mountains and valleys, landscapes, and seascapes.[51] There is a sharp demarcation between honest contemplation of nature and the pathological indulgence in novelty. Curiosity is a disease. The demarcation line is ambiguous, but a line has to be drawn. The sense of sight must be healthy, for otherwise vision will be distorted. But health of the senses is a delicate concept. When we see through a colored glass we see colored things. This sort of deception is well known to everybody, an artificially induced disease of the sense, and yet there are other deceptions that are more subtle. For instance, melancholy is a complexion that may generate pestilential humors, and when people have a strong imagination they can easily mistake what they imagine for the true object perceived with the senses.[52]

Thus, we must pause here and wonder with Galileo under which circumstances the Jovian planets can be certified to be truly existent, and the possibility can be ruled out of their being simply figments of a melancholic imagination, or other complexion that may affect vision. The interposition of glasses between the object and the eye is automatically susceptible to the

48 "La lingua (come dice S. Iacomo) è peggiore di qual si voglia fiera bestia, perche le fiere, ancorche che siano serpenti, con l'ingegno dell' huomo si domano, et si sottopongono, ma la lingua niuna forza humana puote sottoporre; virtù, et forza ha da essere dal cielo, per ciò fare" (Arias 1602, II, p. 244).

49 Arias 1602, II, p. 211.

50 "È disordine della vista quando si desidera vedere cose segnalate, et mirabili, come di edifizii, città, vasi d'oro, et d'argento, animali, giardini, et se desidera questo solo per lo diletto della vista senz'altro fine buono. Et ancora quando desidera di vedere cose curiose [...] o vedere un mostro, et si pretende solamente sodisfare con quel vedere al desiderio di sapere, che cosa sia, o al desiderio di vedere cose nuove" (Arias 1602, II, p. 217).

51 Arias 1602, II, p. 217.

52 Arias 1602, II, p. 220.

risk of artificially inducing a disease of the sense of sight that might lead the observer to indulging in self-deception. There are dangerous consequences of the artificial conditions of vision through the *perspicillum* that may disqualify its outcome as a basis for sound judgment. Worse, artificiality may solicit the unhealthy tendency to pursue curiosity for the sake of curiosity, whereas mortification of curiosity is the precondition for reasoning well.

Finally, we have come to the crux of mortification as central to the reconstruction of the self in the framework of early modern Christian spirituality. It is the mortification of the intellect, or judgment. Personal judgment can be deceiving. To mortify one's judgment means to seek the opinions of others in order to submit one's judgment to the evaluation of other people.[53] Obviously the mortification of judgment need not be applied in cases of truths that must be believed because they are absolutely certain, such as those obtained by divine revelation. But in all other cases, when we believe in things because our judgment tells us that they are such and such, and because a state of affairs seems to be perfectly true to us, we must proceed very carefully and mortify our judgment, especially if we see that other persons equally well informed and who might have experience and knowledge do believe otherwise.[54] We must recognize the state of profound ignorance in which we find ourselves after the fall. The natural light of reason which we have by nature has not been dimmed by original sin, but it teaches us only a few truths with clarity and certainty.[55] It is doubtful whether the existence of the Jovian planets might be a truth that escapes the discipline of the mortification of judgment. The natural light of reason cannot teach us this cosmological truth. The *perspicillum* must condition his sense of sight in order for Galileo to be able to observe the tiny spots of light in the vicinity of Jupiter. But to restore the self in its capacity for judgment in the aftermath of the fall of Adam calls for uncompromising mortification of both the intellect and the senses, and above all for the mortification of speech. It calls for silence.

---

53 "Questo giudizio si ha da mortificare non fidandosi disordinatamente di lui, ma sottoponendolo al giudizio, et al parere de gli altri in tutto quello che sarà conforme alla ragione"(Arias 1602, II, p. 200).

54 Arias 1602, II, p. 200.

55 Arias 1602, II, p. 201.

A Hermetic Parody of the Fall of Man

The spiritual exercise of silence, the therapeutics of *mortificatio*, was accentuated by poetic revivification. Galileo amplified Andrea Salvadori's draft poem for the Medicean stars by offering a hermetic parody of the Christian myth of the Fall of Man. I suggest that Galileo's poetic labor be read as a *distentio animi* (temporal manifestation of the soul) due to a tension of opposites enforced by Christianity's emphasis on the deterioration of physical and mental powers engendered within each individual by original sin.

The poem tells the story of the giants, sons of the earth, unsuccessfully assaulting the throne of Jove, being exterminated, and cast to the kingdom of Hades. In the last three stanzas it reveals its intent of denouncing the grotesque attempt by Galileo's detractors to steal from Jupiter the newly discovered Medicean stars. The poem is a labyrinthine peregrination around a metaphoric centre propelled by the art of *concettismo*. The pagan myth was recounted briefly in the first book of Ovid's *Metamorphoses*. It appeared in a Christianized framework in *Le metamorfosi di Ovidio, ridotte da Giovanni Andrea dell'Anguillara in ottava rima*, which, in the Bernardo Giunti edition of 1584, was illustrated by a series of dramatic engravings produced by Iacopo Franco.[56] In the scene accompanying the first book (fig. 5.2), the Christian God presides over the creation stories recounted by Ovid. The giants are defeated by Jupiter, riding an eagle (a prototype concept of military aviation), under the supervision of the Christian God sitting higher above among the stars, sided by the two luminaries.

56 Anguillara 1584, pp. 1 ff.

Fig. 5.2. A Christianized representation of the myth of the giants assaulting Jove, from *Le metamorfosi di Ovidio, ridotte da Giovanni Andrea dell'Anguillara in ottava rima* (Anguillara 1584, facing p. 1). From: Getty Research Institute, Los Angeles (2959–447), in the public domain at Internet Archive (https://archive.org/.).

Thus Galileo, the natural philosopher in the Garden of Eden, trained the *perspicillum* to the heavens in order to seek the four companions of Jupiter. The instrument prolonged his eyesight, the visual ray issuing from the eye, to penetrate the crystal spheres up to the throne of Jove and seize knowledge of the four moons. Then Galileo, the sidereal messenger, now sublimated into the spirit Mercurius, descended from heaven to announce the celestial spectacles long hidden from view to all of us ordinary mortals. He was assumed into the pagan pantheon in apotheosis. The progeny of earthly philosophers, the giants shouting their bestial ignorance against Jove, was annihilated with a vengeance by Jove's lightning, unlike Adam's descendants who were not condemned forevermore but given a chance to reunite with God by grace and good works (at least in post-Tridentine Catholicism). The state of original justice enjoyed by Galileo in the Garden of Eden, which clothed his human nature supernaturally, was not lost following his infraction for tasting the fruit of the tree of knowledge of astronomy. Instead he was rewarded by the gods.

It is the pugnacious ignorance of the philosophers, the arrogant sons of mother earth, that forever damns them to hell. Philosophers are naked beasts. They suffer from the wounded cognition caused by original sin, having lost the supernatural gifts that were the endowment of the first humans in the Garden of Eden. The supernatural gifts clothed human nature, and man is naked after the fall. The striking metaphor is Roberto Bellarmino's.[57] It visualizes the corruption of human nature as an aboriginal nakedness. Bellarmino states that after the fall of Adam human nature is not any more different from its state of original justice (as far as purely natural things are concerned) than a naked person is from an undressed one. Neither is fallen human nature more deteriorated or laboring under more ignorance than it would be if it had been established in its own natural state.[58] The wounds of fallen mind are ten according to Bellarmino. One is difficulty in understanding the nature of

57 "…non magis differt status hominis post lapsum Adae a statu eiusdem in puris naturalibus, quam differat spoliatus a nudo" (Bellarmino 1858, p. 24).

58 Bellarmino 1858, p. 24. There is a pregnant tension of opposites here between Bellarmino's defense of the Catholic interpretation of original justice as supernatural gift of God supervening on Adam's incorrupt nature and the metaphor of corruption as falling into a nakedness that is both equal to Adam's nakedness *in naturalibus* and yet forever suffering from, hence corrupted precisely because of, the wounds ensuing from original sin. The tension must become even more intolerable for Catholics when contrasted with the sacrament of Baptism that supposedly washes away the blemishes of all sins and forgives all punishments, and yet there perdure in fallen humans temporal consequences of, or punishments for, the sin of Adam.

things. Human mind is so corrupt that it cannot understand even the nature of an ant. Even worse is ignorance of one's own mind. Angelic minds, says Bellarmino, know themselves perfectly, but fallen human mind knows nothing of itself, for it ignores not only its own nature but its very operation.[59]

Though vision with the naked eye is enfeebled by the wounds of fallen mind, the eye aided by the *perspicillum* turns Galileo's vision into a supernatural mode of knowing, an angelic mind who knows its operation. Hermeticism teaches humans about companionship with God, and Galileo's parody, the *distentio animi* born of *mortificatio* of the spirit as a cure for the corruption of human cognition, mitigates the tension of opposites generated by looking over the abyss of original sin.

59 Bellarmino 1858, pp. 248–249.

· 6 ·

HEAVEN

The mathematization of nature has often been associated by historians and philosophers of science with a Platonic world view. As far as Galileo's science is concerned one might assert that the interpretation (and, to a certain extent, the myth) of a Galileo's Platonism was inaugurated very early on in the seventeenth century. Carlo Roberto Dati, an eminent figure within the circle of Galileo's disciples and later on a member of the *Accademia del Cimento*, wrote a *Dissertation* on the utility and pleasure of geometry, in which he reported a dialogue between Galileo, an unnamed young nobleman, and Braccio Manetti, a Florentine gentleman. In Dati's *Dissertation*, Galileo is portrayed as a Platonic philosopher praising geometry for allowing us not only to penetrate the mysterious fabric of the visible universe, but also to transcend its physical appearance by reaching out beyond the stars.[1] It is outside the scope of the present research to reconstruct the complex historiographic development of the various interpretations of Galileo's attitudes towards Plato, a theme that would doubtless deserve further investigation. Galileo was interested

1 Carlo Roberto Dati was a scientist, a linguist, and a literary theorist. On his *Disseration*, cf. Garin 1978, II, 848–849. I have read Dati's *Dissertation* in Targioni Tozzetti 1780, II, 314–327. The dialogue narrated in the *Dissertation* would purportedly have occurred at Galileo's villa in Arcetri when Dati was in his youth.

in methodological questions related to the maieutic mode of inquiry, which appear to be related to certain performative aspects of hermeticism that I have already touched upon. Galileo's awareness of the potential of mathematical reasoning for natural-philosophical inquiry seems to have undergone a significant anti-Platonic development (by the standards of late sixteenth-century Platonism). He started out in the early 1590s as a keen supporter of the role of mathematics in natural philosophy, and he might have subscribed to a philosophical method privileging some form of maieutics, inspired by the Platonic teaching of the then famous Jacopo Mazzoni, one of his masters at Pisa University. But by the time he published the *Dialogue on the two chief world systems* (1632), he had turned to a new point of view that conferred epistemological primacy on mathematical reasoning and cast serious doubt on the validity of maieutics and the theory of reminiscence. I would like to pursue this theme of the performative dimension of hermeticism by reconsidering Galileo's relation to Plato's philosophy.

Galileo and Plato

Concluding his *Études Galiléennes* (1939), Alexandre Koyré put forward his seminal thesis concerning Galileo's Platonism, portraying Galileo's mathematization of nature as the vindication of a Platonic approach to the study of the natural world. Koyré further articulated his view on Galileo's Platonic science in a paper published a few years later.[2] First of all, in Koyré's view, the "role of mathematics in physical science was not a new problem [i.e., at the time of Galileo]. On the contrary, it had been the subject of philosophical meditation and debate for two thousand years: and Galileo was well aware of this".[3] Koyré adduces the discussions Galileo must have had with the Aristotelian Francesco Buonamici and the Platonic Jacopo Mazzoni at Pisa University as evidence that Galileo knew very well the historical pedigree of the problem of mathematical argumentation in natural-philosophical inquiry. Buonamici and Mazzoni, Koyré argues, simply represented the average view present in the science and philosophy of the time, according to which "there was a sharp dividing line between Aristotelians and Platonists".[4] In other words, those

2 My quotations from Koyré's *Études Galiléennes* are from Koyré 1978. The later paper I am referring to is *Galileo and Plato* (Koyré 1943).

3 Koyré 1978, p. 201.

4 Koyré 1978, p. 201.

"who attributed to mathematics a higher value, and who in addition saw it as having real value and a dominant position in and for physics, were Platonist: those who, on the contrary, saw mathematics as an 'abstract' science, and consequently as of lower value than the sciences (physics and metaphysics) which were concerned with the real, and in particular those who claimed to base physics directly on experience and attributed to mathematics only a subordinate role, were Aristotelian".[5] Koyré, however, stressed that it was not so much the epistemological problem of certainty that was in question, as the ontological problem of reality. He admitted that no Aristotelian would have denied "the certainty of geometrical proof', and therefore, for him, it was simply 'a matter of the role of mathematics in and for the very structure of scientific knowledge". Koyré declared that he was in agreement with Galileo's contemporaries' epistemological conception. According to him, "the mathematization of physics is Platonism…Therefore, from our vantage point we can see that the advent of classical science was a return to Plato". In the context of late sixteenth-century philosophical culture, Koyré saw the Aristotelian refusal to recognize the role of mathematics in the investigation of nature as a perfectly plausible attitude towards the study of physics. In Koyré's words, the Aristotelian attitude "seems perfectly reasonable, and the objections that Aristotle had addressed to Plato cannot be refuted, except in practice. For one cannot logically prove something to be possible. The only proof that something is possible is that it exists. So in order to prove that it is possible to formulate exact mathematical laws which apply to reality it is necessary to actually formulate them".[6] Galileo, Koyré continues, was perfectly aware of this and it was only by furnishing a concrete treatment of mathematical physics that he could prove the Aristotelians wrong. The Aristotelian of Galileo's *Dialogue*, Simplicio, argued that "mathematicians may prove well enough in theory that *sphaera tangit planum in puncto*…but when it comes to matter, things happen otherwise". In Koyré's opinion, "from the point of view of Aristotelianism, and even from that of ancient Platonism, it [i.e., Simplicio's argument] is decisive and irrefutable. For in the real world, the physical world, there are no straight lines, nor planes, nor triangles, nor spheres". Thus, there is inevitably a divergence between "geometrical figures and real shapes", so that between "the essence and its realization there is always some *play*".[7]

5 Koyré 1978, pp. 201–202.
6 Koyré 1978, pp. 202–203.
7 Koyré 1978, p. 203.

In 1972, William Shea developed Koyré's thesis, by further articulating the conclusive remarks concerning the Socratic method that Koyré had made in the final part of *Galileo and Plato*.[8] In Shea's view, "Plato claimed that the soul, before it enters the body, is directly acquainted with the Forms upon which the world is modeled. When it enters the body all this calm knowledge is drowned in the flood of new sensations. But although it is all forgotten, it is not destroyed. Later experience of things modeled after a certain Form may jog one into recollecting one's prior knowledge of the Form itself. Experience, on this view, is indispensable for arriving at knowledge which transcends experience. Galileo does not say that the soul existed before the body. But he does say that the human mind is pre-attuned to nature".[9] According to Shea, Galileo fully endorsed the Socratic method and Salviati in the *Dialogue* practised 'mental midwifery'. Referring to the maieutic-mathematical episode, in which Simplicio asks Salviati whether he embraces Plato's theory that *nostrum scire sit quoddam reminisci*, Shea points out that Galileo is not entirely explicit in his answer, a fact that escaped Koyré's notice. "It would be nice", Shea observes, "to be able to report that Salviati replied, *Four-square with Plato*. But at this point Galileo's epistemological caginess again reasserts itself. Salviati replies that he has already indicated what he felt about Plato's opinion by deeds rather than by words, and that he will continue to follow this method".[10] Thus, Shea concludes, Galileo "agreed with Plato that knowledge of at least some natural principles is self-evident, and, in a way, innate, but he stressed the importance of experience to awaken dormant knowledge. Ultimately, this is possible because the human mind shares in the divine mathematical knowledge that created and ordered the universe. In a sense, there exists between man's intellect and nature a pre-established harmony which experiments can recall rather than produce".[11]

However, during the six decades since it was first formulated Koyré's thesis on Galileo's Platonism has not remained unchallenged, both on philosophical and historical grounds. One might draw the conclusion that at least one aspect of Koyré's thesis depends on a highly problematic vision of Platonism

8 Shea 1972, pp. 150–155. In reading the conclusions of Koyré's later essay, *Galileo and Plato*, one has the impression that he somehow attenuated his previous interpretation. The paper ends emphasizing the significance of Galileo's apparent endorsement of the Socratic method based on maieutics (Koyré 1943, pp. 427–428).

9 Shea 1972, p. 151.

10 Shea 1972, pp. 151–154.

11 Shea 1972, p. 155.

and Aristotelianism, that is, the former as the warrant of the mathematization of nature and the latter as the very antithesis of it, as two rather abstract and a-historical categories.[12] In fact Platonism and Aristotelianism should be seen at most as broad interpretative schemas that must be attuned to the historical variety of settings within which, in the sixteenth century, those two philosophies gradually mutated into various strains. It might be added that in order to validate his thesis on the divide between Aristotelians and Platonists on the question of mathematics, Koyré saw fit to pair the passage from Francesco Buonamici's *De motu* with another one from Jacopo Mazzoni's *In universam Platonis, et Aristotelis philosophiam praeludia*, a book by another philosophy professor at Pisa, lecturer *extra ordinem* on Plato from 1589 to 1597, friend, teacher, and colleague of Galileo.[13] In this second case, Koyré simply mistook a hint at the beginning of Mazzoni's discussion for another proof that such an intellectual cleavage had actually existed. Mazzoni wanted to draw upon the fact that the question of mathematics had given rise to many speculations in order to create the context in which he himself, by direct comparison of both Aristotle and Plato, was to collect the truth on various specific issues concerning natural philosophy. Behind the mask of its apparently unprepossessing title, Mazzoni's book presented a much more ambitious program of inquiry that went far beyond the schematic vision underlying Koyré's thesis. That program extended far beyond the Koyréan question of a supposed philosophical chasm dividing Plato and Aristotle over the role of mathematics in natural philosophy.[14]

---

12 As early as 1957, Eugenio Garin pointed out a number of reservations about the way in which Koyré had treated the question of an alleged Platonism in Galileo (Garin 1957).

13 Koyré 1978, p. 201 and Koyré 1943, p. 421.

14 "As the hunter painstakingly looks for his animals, which hide in forests and thickets, and after capture enjoys the game and shares it with others, so he who is looking for learning makes every effort to ferret out truth from dense matter and opaque sensile involucra, in order to have fruition of it and share it with others. But as hunters are provided with great help by their hounds, without which they could hardly do anything, so he who is learning needs his own hounds to accomplish his pursuit. These hounds (as Plato clearly teaches) are the ancient sages who bark at ignorance and coax us into science and learning. Not only do they show us the road towards truth, they teach us how to capture truth by securing it with the knots of demonstrations. Thus, following this advice, to hunt through the vast territory of being, I have taken advantage of two great hounds, Plato and Aristotle, who, more than anyone else, are used to capturing truth and opening it up with solid and subtle arguments, so that its most profound mysteries, like the bowels of the earth, become clearly visible" (Mazzoni 1597, *Dedication* (not paginated)).

In conclusion, there has been in the literature a tendency to recognize the vital element of a Platonic *Weltanschauung* in Galileo's approach to the mathematization of nature, in terms both of the metaphysical relationship between knowledge and the natural world and of the method with which that knowledge is to be pursued. More precisely, whereas Koyré has pointed out that Galileo's ontology, though inspired by Plato, went well beyond Plato in staking a claim to the homogeneity of the real and the geometrical, Shea has emphasized Galileo's rather ambiguous endorsement of Socratic maieutics. Rather than seeing a sharp dividing line, we should think more in terms of a blurred distinction. This leads us again to the connection between Platonism and performative hermeticism. But first we need to meet Galileo's Platonic teacher.

Jacopo Mazzoni, Teacher of Galileo

I will look at how Mazzoni tackled the role of mathematics in natural-philosophical inquiry, within the context of his broader Platonic conception of philosophy as an enterprise aiming at the knowledge of God through a preparatory exercise, which he called 'clearing' (*purgatio*).[15] This presentation is based on what should probably be regarded as Mazzoni's masterpiece, i.e., his book *In universam Platonis, et Aristotelis philosophiam praeludia*, published a year before his death. He devoted an entire chapter of his book to an analysis of what he considered the Platonic 'clearing', a progression through a series of propaedeutic stages, according to a method whose final goal was the philosopher's attainment of the contemplation of pure forms, and ultimately of God.[16] The process of clearing, in Mazzoni's view, passes through three stages. First,

15 Mazzoni 1597, pp. 142ff. Literature on Mazzoni's philosophical method is scant. Purnell 1972 is mainly concerned with Mazzoni's philosophical eclecticism and his attempt at reconciling Plato and Aristotle, especially on the question of the applicability of mathematics to physics.

16 Mazzoni 1597, pp. 142–152. It is worth noting that Mazzoni tends to group together a huge number of Ancient and Christian sources under the general heading of 'Platonic'. Thus when he refers to Plato and Platonism one has always to keep in mind that his notion of Platonism was much broader than that which modern scholars would presumably be prepared to accept. This is another indication of the limitations of Koyré's conception of Platonism. For example, on p. 144, Mazzoni quotes *Sophist* 227C–228E, as a locus which would, for him, put forward very clearly the necessity of a preparatory clearing of the mind. Yet, a few lines after citing Plato, he immediately goes on to quote Lactantius, from the *Divinarum institutionum*, who, Mazzoni claims, has best of all expressed that necessity.

there is the clearing removing impediments (*purgatio impedimenta removens*). Second, there is the clearing directing (*purgatio dirigens*). Finally, there is the preparatory clearing (*purgatio preparatoria*). The first stage refers mainly to a moral and didactic dimension. It amounts to correct education (*recta paedia*) and aims at inculcating in the young the conviction that honest labor and suffering are to be accepted and given preference over alluring pleasure. Gymnastics and music are for Mazzoni valuable means to that end. The second stage is called 'clearing directing' because, Mazzoni argues, through such kind of clearing our intellect is ultimately directed to instituting its relationship with God.[17] The correct name for this second stage of the clearing process, Mazzoni suggests, is 'dialectics'. Referring to *Philebus* (probably 15 E), Mazzoni describes Plato's conception of dialectics with a simile, according to which the dialectician is the philosopher who is able to emulate God's Titanic powers to constitute single things as though tearing Osiris into shreds (*quasi Osirim in multitudinem vi Titanica discerpens res constituit*) and subsequently to attract the single things back towards himself, transforming and perfecting them (*…res ad se retrahit, convertit, et perficit*).[18] Since, according to Mazzoni, this approach confers on dialectics such a force, he feels it necessary to explain why the name 'dialectics' is precisely the appropriate one, instead of, for instance, that of 'metaphysics'. His argument is as follows. All things that are constituted by two different things, such as form and matter, take their name from the form. Thus, since metaphysics is constituted by the knowable things (contributing matter) and by dialectics itself (contributing form, i.e. order and method), then it is only too appropriate to call the supreme form of knowledge offered by metaphysics 'dialectics'. Mazzoni is keen to distinguish dialectics from all other particular sciences, such as mathematics, which, in his opinion, is unable to aspire to the status of most general science of all.[19] The third and final stage of clearing is called 'preparatory' because, just as those who have long stayed in the darkness need a period of transition during which they can adapt to dim light before being exposed to the brightness of the sun, so those whose soul has long been obtenebrated by ignorance need first be exercised in shadows [umbrae]. From the shadows of bodies [corpora], the learners may

17 Mazzoni 1597, pp. 148–149.

18 Mazzoni 1597, p. 148. This is a reference to the myth of Osiris killed and cut to pieces by his brother Typhon.

19 Mazzoni 1597, p. 149.

ascend to the bodies themselves, from these to the *rationes* hidden within their soul, and finally upwards to the Ideas.[20]

In order to reach a better understanding of exactly where Mazzoni places mathematics within the process of clearing, we need to focus on his interpretation of what he refers to as 'the Platonic order of things'. Quoting Proclus's commentary on the *Republic*, Mazzoni maintains that the order of all worldly things (*mundane*) can be compared to a line and divided up according to a number of genera. Thus, Mazzoni claims, given the orderly structure of the world it should not come as a surprise if our mind is able to progress from the properties of shadows up to the properties of bodies. But given that how to achieve this passage, according to him, has never been taught by any Platonist, he goes on to show how, by means of mathematical reasoning, one can effectively transit from the knowledge of shadows to the knowledge of material bodies.[21] Finally, Mazzoni discusses whether and to what extent mathematical reasoning can really be used in physical matters. He devotes an entire section to this delicate question, in which he repeatedly castigates Aristotle for not having understood the value of mathematical proofs on too many important occasions.[22] In essence, for Mazzoni, true philosophy can only be that most fundamental way of knowing which eventually connects us with the realm of divine forms. All that precedes that very final stage is simply the gradual process of clearing, a propaedeutic sequence of purificatory steps towards the attainment of the knowledge of God. However, Mazzoni does not clarify what the method of clearing really consist of, especially from a practical point of view. Is it a mental gymnastics entailing the presence of a teacher acting as an intellectual midwife in the style of Plato's Socrates? Or is it simply intended as a solitary exercise? Furthermore, and most importantly for our purposes, is mathematics simply a part of the process of clearing, does it intervene after clearing has been achieved, or does it play a double role? Especially on the latter point, Mazzoni, rather disappointingly, is none too

20 Mazzoni 1597, 149. Here I cannot expand on Mazzoni's metaphysics and his conception of reality as constituted by degrees of being whose ultimate guarantee is God. Cf. Purnell 1972.

21 It must be pointed out that Mazzoni had some knowledge of the mathematics of his time. He certainly grasped the elementary aspects of proportion theory, though he seems to have been unable to produce any original mathematical arguments (cf., for example, Mazzoni 1597, p. 150, pp. 181–183, and 190ff.). However, the mathematical example adduced by Mazzoni, based on the idea of computing the height of a body with a simple proportion by measuring the length of its shadow seems rather naïve (Mazzoni 1597, pp. 149–150).

22 Mazzoni 1597, pp. 187–197.

perspicuous. It is true that in his view Aristotle's fault was not so much to have rejected mathematics wholesale as to have failed to recognize its import on a number of very specific questions.[23] I would argue that Mazzoni recognized the powerfulness of mathematics in solving topical puzzles in Renaissance natural philosophy.[24] But, as we have seen, he clearly saw dialectics as far superior to mathematics, which he considered just a particular science. To complicate the picture further, at a later point in his book, Mazzoni distinguishes dialectics from clearing, whereas he previously equated the former with the second stage of the latter. He asserts that "Plato knew only one science, which he called dialectics, and to which he ascended via the triple process of clearing".[25] Thus, Mazzoni may have oscillated as to the relationship between dialectics and clearing, but he was obviously unwilling to elevate mathematics to the higher epistemological status that he reserved for dialectics.

But what about maieutics and the theory of knowledge as reminiscence? Whereas the first stage of the clearing process refers to education, and thus implicitly seems to involve the active role of the educator, the second and the third stages remain more ambiguous as to their practical implications. Mazzoni subscribed to the immanentism of the *rationes* of all things within the rational soul and also to a general theory of knowledge as reminiscence.[26] To the extent that we are willing to accept that the theory of reminiscence presupposes some form of maieutics, we can assert that Mazzoni automatically endorsed maieutics. Unfortunately he does not explicitly discuss maieutics in relation to clearing. On this point, therefore, I believe that we have to accept an indeterminacy, but I think that it is reasonable to hypothesize that Mazzoni would have fully supported the Socratic method (there seems to be nothing in his book suggesting the opposite).

Finally, we must note that, in connection with his conception of propaedeutic clearing and maieutics, an interesting question arises as to the characteristics of the style of teaching practiced by Mazzoni at Pisa. According to an item of information by Pier Segni, who recited the funeral oration for Mazzoni, we know that the latter was amicably courted not only by students

---

23 Mazzoni 1597, p. 190.

24 Amongst those puzzles, there were, for example, Aristotle's theory of the impossibility of motion in a void space, the problem of the weight of bodies within the elements of air and water, and the question of the mobility of pyramidal shapes, which, according to Mazzoni, Plato had assigned to the element of fire.

25 Mazzoni 1597, p. 179.

26 Mazzoni 1597, p. 175.

from all faculties, but also by other professors. Segni also tells us that Mazzoni used to teach, guide, and answer his friends 'with great love', so that, according to Segni, everybody was fully satisfied with Mazzoni's intellectual and moral manners.[27] On 15 November 1590, Galileo, already a professor of mathematics at Pisa, wrote his father Vincenzio that he was very well and was learning from Signor Mazzoni.[28] A month later, Guido Ubaldo dal Monte, the patron who had most probably brokered Galileo's appointment at Pisa, answering a now lost letter by Galileo, asserted that he was very pleased to know that Galileo and Mazzoni were enjoying friendly discussions (*ragionamenti*), in which he regretted that he could not take part.[29] Does this fragmentary evidence suggest that Mazzoni's mode of teaching somehow reflected his ideas on the method of clearing and maieutics? Obviously, we do not have a decisive argument to answer this question positively, but we might venture to hypothesize that Mazzoni was indeed practicing something of the method of clearing as a way of preparing his friends and students to philosophy, perhaps in a fully-fledged Socratic manner, by questioning them and eliciting their answers. It is a possibility that is not inconsistent with the scant biographic data mentioned above, which signal a certain aura of fascination with the personage. We shall see that Galileo's response to an anti-Copernican argument put forward by Mazzoni in his book can to all intents and purposes be read as an exercise in the application of the clearing, implicitly challenging his old master's philosophical assumptions.

Extemporization in Galileo's Letter to Mazzoni

Galileo wrote to Mazzoni on 30 May 1597 from Padua, immediately after seeing the latter's book on Plato and Aristotle. Given the paucity of evidence explicitly concerning Galileo's early Platonism offered by the surviving documents, it is essential to read the very few elements we have in their appropriate context. The most important document relevant to Galileo's early Platonism is the letter to Mazzoni, not only because it was written to a former tutor who also happened to be a Platonist, but also because its interpretation requires the reconstruction of a complex historical context. One of the most intriguing problems related to the Mazzoni letter is this. The letter gravitates around a

---

27 Serassi 1790, pp. 90–91.
28 Galilei 1890–1909, X, pp. 44–45.
29 Galilei 1890–1909, X, p. 45.

sophisticated mathematical counter-argument to an anti-Copernican objection raised by Mazzoni. The apparently robust nature of Galileo's rebuff notwithstanding, that counter-argument surprisingly never made it into his later *Dialogue*, where Galileo claimed to have marshaled all the argument against Copernicanism. It simply disappeared. Why? William Shea has studied the mathematical structure of Galileo's counter-argument, but neither he nor other historians have tried to answer this question.[30] The counter-argument is undoubtedly in favor of Copernicus, but against whom did Galileo write it? An answer to the latter question is not as straightforward as might appear at first glance. Instinctively, Mazzoni would seem to have been Galileo's obvious target. However, the general tone of the letter is by no means confrontational and hostile. Its contents and style rather suggest a number of indirect references to Mazzoni's Platonism. Even more puzzlingly, some three months after writing to Mazzoni, Galileo answered Kepler thanking him for sending the *Mysterium cosmographicum* and claiming not only to have long been a Copernican, but also to have written a number of papers both to disprove anti-Copernican arguments and to explain various phenomena in accord with Copernicus's system. He also lamented that he had been unable to publish those works because of unfavorable circumstances.[31] Somewhat predictably, historians have had sneaking suspicions about Galileo's claims, and indeed, apart from the mathematical counter-argument in the Mazzoni letter, there is nothing in the historical record to suggest that the claims are founded. Galileo, it would seem, simply made a resounding and rather empty boast. He would have to wait for some thirteen years and the arrival of the telescope before finding new evidence for the Copernican system. In order to look for a possible explanation of all these disparate elements, of the baffling demise of Galileo's counter-argument, of the question of the apparent lack of a polemical target, and first and foremost of the Platonic references concealed within the letter, I propose that we turn to some aspects of the style of performance that, as we have seen, was an element of hermeticism.

Anthropological and semiotic studies suggest that various patterns of socio-cultural behavior might be interpreted as instances of performance. Performance is "a dimension of human life that is most typically emphasized in music, theatre and other public displays of artistic abilities and creativity".[32]

30 Shea 1972, pp. 111–113.

31 Galilei 1890–1909, X, p. 68.

32 Duranti 1997, p. 15. Cf. also Hanks 1989, pp. 111–112, and Palmer and Jankowiak 1996, p. 226.

But the dimension of performance is also found in "verbal debates, story tellings, singing, and other speech activities in which what speakers say is evaluated according to aesthetic canons, that is, for the beauty of their phrasing or delivery, or according to the effect it has on the audience, namely, for their ability to "move" the audience".[33] I am particularly interested in a few analytical developments related to research on performance. These developments refer to: a) the general notion of indexicality; b) the relationship between performance, composition, and improvisation; c) the process of entextualization and its relationship with performance; d) the notions of the emergent and improvisational creativity. I will illustrate them briefly, before applying them to the analysis of the episode documented by the Mazzoni letter. Indexicality is basically a Peircean notion. "An index is a sign which requires an association between the sign and its object. The classic example of an index is the weathervane, which indicates wind direction".[34] An index is to be distinguished from a symbol, of which "each token, or instance is connected to its object only by an underlying or immanent general rule and nothing else". The index is "also not isomorphic to its object, thus distinguishing it from the icon, which is connected to its object by similarity of sign-autonomous properties".[35] The notion of indexicality has been applied, for example, in pragmatics, especially to deictics, signs whose referent depends on the indexical relation that a token has to the context of usage (e.g., 'this'). Indexical presupposition refers to the fact that a certain topic currently in play during a verbal interaction has already been identified as being that of the surrounding stretch of discourse. Indexical entailment represents an indexical relationship "to some presumed future direction of the interaction".[36]

Drawing on a comparative study of musical performances in different cultures, Bruno Nettl advanced the hypothesis that improvisation and composition are not opposed concepts, and suggested that we should not juxtapose them, but rather see improvisation and composition as part of the same idea.[37] Keith Sawyer proposed that instead of considering composition and improvisation as being the ends of a spectrum, different types of performance should

33 Duranti 1997, p. 16.

34 Sawyer 1996, p. 274.

35 Sawyer 1996, p. 274.

36 Sawyer 1996, p. 275.

37 Nettl 1974, 6. Also in the Western musical tradition, Nettl suggests, we can see that the polarity composition/ improvisation is questionable, "by comparing, for example, the painstaking and often protracted method of Beethoven with Schubert's quick, spontaneous creation of lieder" (*ibid.*).

be contrasted along nine different dimensions, so that composition and improvisation represent "points maximally opposed in a multidimensional space" (cf. Table 6.1, to be discussed in a moment).[38]

Table 6.1. The nine contrast dimensions of Keith Sawyer's multidimensional space

	Ritualized/Entextualized[39]	**Improvisational**
1	Ossification	Revivalism
2	Low creative involvement	High creative involvement
3	Indexically reflexive	Indexically entailing
4	Narrow genre definition	Broad genre definition
5	Large ready-mades	Small ready-mades
6	Low audience involvement	High audience involvement
7	Resistant to novelty	Receptive to novelty
8	Changes long-lasting	Changes short-lived
9	High cultural valuation	Low valuation

According to Sawyer, the maximally opposed points of the multidimensional continuum are more appropriately considered to be 'ritualized performance' and 'improvisational performance'.[40] This means that improvisation is not an on/off quality and that we should rather think in terms of a degree of improvisationality, or extemporization, determined by a point in the multidimensional space defined by the nine contrast dimensions. The notion of multidimensional space leads us to introduce the concept of entextualization and its relationship with performance. According to R. Bauman and C. R. Briggs, entextualization is the "process of rendering discourse extractable, of making a stretch of linguistic production into a unit—a text—that can be lifted out of its interactional setting".[41] Saywer has tried to generalize this notion by advancing the hypothesis that a pragmatic, indexical approach to creativity suggests the range (not intended to be exhaustive) of contrast dimensions described in Table 6.1. In this more general context, entextualization would be the process by means of which a performance with a high degree of improvisionality (a point located towards the categories presented

38 Sawyer 1996, p. 273.

39 Sawyer's table proposes 'ritualized' instead of 'entextualized' (cf. the reasons for that choice in Saweyr 1996, p. 282). For our present purposes, the terminological preference expressed by Sawyer is not important and we shall keep to 'entextualized/ entextualization', according to Bauman and Briggs's terminology (Bauman and Briggs 1990, p. 73).

40 Sawyer 1996, p. 273.

41 Bauman and Briggs 1990, p. 73.

in the right column of Table 6.1) is transformed into a more stable cultural product, such as, for example, a literary text, a musical score, or a religious rite (a point located towards the ends presented in the left column of Table 6.1). Dimensions 2–4 and 6–8 are particularly relevant for the approach adopted here.[42] Dimension 2). A higher degree of improvisationality is related to a higher creative involvement on the part of participants in improvisational genres, such as conversation and debate, especially through the means of indexical entailment. Dimension 3). Reflexive indexicality is typical of poetic means of expression, in that poetic structures tend to constrain the possible audience interpretations of the performance. On the other hand, at each moment of a more improvised performance, a range of possible indexical entailments provides for the impromptu creativity emerging in the process. Dimension 4). This contrast refers to the observed fact that in a more improvised genre "a greater variety of performances will be considered 'the same', or tokens of the same performative event type (culturally defined)".[43] Dimension 6). Sawyer suggests that the degree of improvisationality is generally proportional to audience involvement, so that in more improvised genres, "performers are more aware of themselves as responding to an audience".[44] Dimensions 7, 8). More improvisational genres are more receptive to novelty, and innovation is not "likely to last long before later changes push it out of the genre".[45] Finally, inspired by George Herbert Mead's philosophy, Sawyer defines the emergent as the result of creative interaction during performance. The emergent is "structured, but ephemeral, changes with each performance act, and emerges from the indexical presuppositions accumulated through the prior collective interaction".[46] Thus, for Sawyer, improvisational creativity can be characterized as follows.

> Performers are constrained to operate within the performance genre. A given act is more narrowly constrained by the emergent. The nature of this constraint is unique

---

42 *Ossification* vs. *revivalism* refers to the study of the permanence and change of cultural forms, particularly ritual performance. *Large ready-mades* vs. *low ready-mades*: this contrast dimension refers to the availability and size of pre-existing units of composition which the performer can make use of. *High cultural valuation* vs. *low valuation* refers to the cross-cultural higher appreciation of non-improvised performances and artifacts (Sawyer 1996, pp. 280ff.). These two dimensions are less interesting for our purposes.

43 Sawyer 1996, p. 284.

44 Sawyer 1996, p. 286.

45 Sawyer 1996, p. 286.

46 Sawyer 1996, p. 279.

> and specific to the performance and the moment of the interaction. In improvisational genres, each performer is expected to contribute something original to the evolving emergent in each act, through the process of indexical entailment. In the choice of indexical entailment, performers are subject to the constraints of the emergent. In response to the performer's action, the other participants evaluate the act, and the subsequent interaction determines to what extent the indexical entailment resulting from the act affects the (still/always evolving) emergent. This 'evaluation' is often immediate and often not consciously goal-directed. A more skilful indexical entailment is more likely to enter the emergent, thus operating with more force on subsequent performance acts.[47]

How do performance and the analytic categories develop in connection with performance research bear on my present approach and on the question of Galileo's early Platonism? Let us try to articulate an answer by looking at the Mazzoni letter in detail. Here is a schematic of the interpretation I wish to pursue.

Firstly, I suggest that we may answer the question of why Galileo's counter-argument disappeared if we consider it to be characterised by the quality of a rather improvised performance, not so much concerned with launching a Copernican programme in earnest, but rather with issuing a challenge on the role of mathematics in Platonism, targeted both at the Platonic Mazzoni and at a broader audience.[48] Thus, we can make direct use of the notion of entextualization and advance the following explanation of the demise of the counter-argument. First of all, I would argue that the Mazzoni letter is by no means an elaborated literary piece. It is just the indexically dense medium of a performance. The performance Galileo staged elaborating on Mazzoni's methodological assumptions (a point located towards the ends in the right column of Table 6.1) did not succeed in allowing him to produce, or rather re-produce a literary text (a point located towards the ends in the left column of Table 6.1), which could later have found its way into the Dialogue. Why so? Because, I suggest, the process of entextualization failed. It failed largely on account of the radically impromptu nature of that performance. In other words, what we really have in the Mazzoni letter are just some elements of textuality, incorporated in the medium of a fleeting mathematical performance.[49]

---

47 Sawyer 1996, p. 279.

48 According to Antonio Favaro, the Mazzoni letter was widely circulated by Galileo amongst his friends (Galilei 1890–1909, II, p. 195). The autograph of the letter has not survived.

49 It is possible to resort to banal explanations, such as, for example, Galileo forgot all about his counter-argument to Mazzoni, or lost his copy of the letter. However, there are objections to trivial explanations such as these. Antonio Favaro, who reconstructed the complex

Secondly, I believe that we can illuminate the reasons behind the Copernican boasts if, according to Shea's analysis, we consider them as the consequence of Galileo's realizing that "he could use mathematics to dismantle the latest argument against the motion of the Earth", so that Mazzoni's objection somewhat abruptly became the spur which, in connection with the arrival of Kepler's book, may have raised Galileo's awareness of the potential of mathematics for cosmology.[50] Unfortunately, the paucity of evidence in the documents related to this period of Galileo's life prevents us from further expanding on this point concerning Copernicanism.[51] I now turn to fleshing out this sketch by looking in detail at the Mazzoni letter while extracting from it the various dimensions of Galileo's hermetic performance, so as to reconstruct the elements of his early attitudes to Plato.

Indexical density is evident throughout the letter, both on the socio-linguistic level and on the mathematical-philosophical level. On the socio-linguistic level, Galileo begins by observing that not only Mazzoni's book is "replete with great doctrine, for which it is about to be admired and appreciated by the public", but also that he is extremely pleased to see that on many questions which he and Mazzoni used to discuss during the first years of their friendship the former master now inclines towards his pupil's opinions.[52] Unsurprisingly, we are not told what those questions were about since Galileo is here making use of indexical presupposition (the past discussions in the friendly Pisan environment…).[53] But there is also indexical entailment (the book is about to be admired…). Subsequently, Galileo goes on to say that he is also pleased to understand that Mazzoni has now solved some tedious personal problems (again, Galileo is making use of indexical presupposition and does not tell us anything about the nature of those problems). For all these reasons, Galileo continues, he has decided to put pen to paper and express his happiness to his friend by having recourse to the only suitable means at his disposal, i.e., his mental quickness (an indexical entailment pointing to the intellectual development of the contents of the letter). On the

history of the Galilean collection of manuscripts preserved in the National Library at Florence noted that the exceptional richness of the collection has to be attributed first and foremost to Galileo's scrupulousness in gathering and preserving his writings (Favaro 1885, pp. 2 ff.).

50 Shea 1972, p. 113.

51 Here, again, it is possible to resort to banal explanations, but, as the comments I already made suggest, I believe that they would be unsatisfactory.

52 Galilei 1890–1909, II, p. 197.

53 Those questions cannot concern Copernicanism since Mazzoni remains anti-Copernican.

mathematical-philosophical level, it must be noted that Galileo claims he is enthusiastic about all the conclusions reached by Mazzoni in his book (note the indexical presupposition, though once again we are not told anything more specific), except the latter's objection to Copernicus. He then hints that he is about to expand precisely on the latter question and overtly suggests that Mazzoni should rebuff his counter-argument, so that he might "stand corrected and change his mind (*mutar pensiero*)".[54] Here, we can observe the presence of both indexical entailment and indexical presupposition on the philosophical level. The hint at the changing of the mind is a reference to Mazzoni's teaching on the clearing process (indexical presupposition). And the invitation to the master to further pursue this point in dialogue with his former tutee is an explicit instance of indexical entailment. Finally, as we shall see, indexicality is unambiguously present both in the form of the technical language chosen by Galileo, proportion theory and practical calculations, and in the form of the diagrams accompanying his proof. Galileo now goes on to quote the essence of Mazzoni's anti-Copernican objection. Note that, with this quotation, Galileo is here artfully determining the extent to which Mazzoni's argument is about to enter into the emergent of the performance. At the same time, by fully accepting Aristotle's authority on the crucial point of the whole argument, Galileo is defining a specific constraint which Mazzoni's response will have to satisfy (indexical entailment).[55] The following is a summary of Galileo's quotation of Mazzoni's argument.

You [Mazzoni] argue that if the sun were at the centre of the cosmos and the earth were orbiting the sun then at midday and midnight we should see different portions of the spherical space of the celestial vault. According to you [Mazzoni], the counter-objection, according to which in the Copernican system the size of the cosmos is so enormous that the distance earth-sun is insufficient to produce any difference in our perception is inconclusive, for the following reason. As we know on the authority of Aristotle (note the constraint indexically posed by Galileo's acceptance of Aristotle's authority on this score), the summit of the Caucasus is illuminated for a long time before the sun's light reaches the base of the mountain. Consequently, from the summit of

54 Galilei 1890–1909, II, p. 198.

55 On indexical constraint satisfaction during performance, cf. Sawyer 1996, 279. Indeed, Mazzoni's argument is more complicated than the account given by Galileo, since Mazzoni is strongly interested in saving Aristotle from some criticisms leveled by certain mathematicians precisely on the central element of the whole argument, i.e. the question of the illumination of the Caucasus (Mazzoni 1597, pp. 196–197).

the Caucasus an observer would see a significantly greater portion of the celestial vault than that visible from the base, and therefore a far greater difference in perception between midday and midnight would have to be observed in the Copernican cosmos, since the distance earth-sun must be much greater than the height of the Caucasus. This, if I understand it correctly, is your [Mazzoni's] demonstration, whose beauty and subtlety cannot be denied.[56]

The general structure of Galileo's mathematical counter-argument, too, is indexically dense and orchestrated on the conceptual notes which form the substance of Mazzoni's objection, i.e., mathematics, the point of view of the observer, the variation of the observer's horizon, the image of a fabulously high mountain, the authority of Aristotle, the image of the sun's light illuminating bodies, and the underlying concept of the cosmos as a sphere. First of all, Galileo asks, is it possible that the great distance between the centre of the earth and its surface cannot cause any difference in the perception of the spherical space, and yet the height of the Caucasus can cause such a significant difference? The solution to this problem, Galileo claims, is the altitude of the point of observation on the surface of the earth.

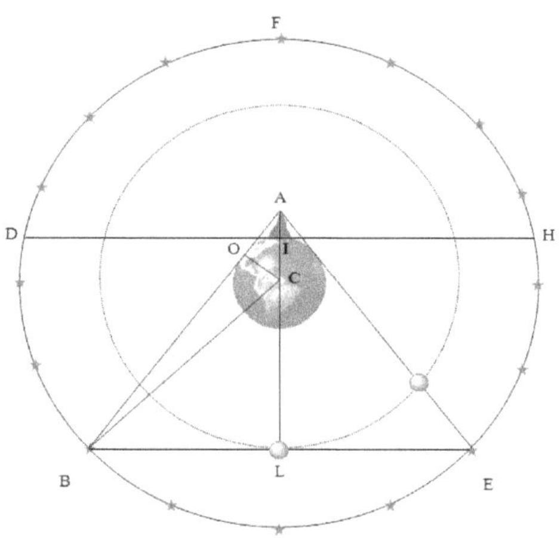

Fig. 6.1. An adaptation of the diagram accompanying Galileo's counter-argument to Mazzoni. The diagram depicts the Ptolemaic cosmos with the earth at its centre. L represents the sun and AI the Caucasus. An observer placed in A would see a portion of the spherical space bounded by the circle of the horizon BE. For the sake of clarity, I have represented the earth and the sun with quasi three-dimensional images, but in Galileo's original the earth is represented by a simple circle and the sun is not represented at all.

56 Mazzoni 1597, pp. 130 ff., and pp. 196–197. See Galileo's reconstruction of Mazzoni's argument in Galilei 1890–1909, II, pp. 198–199.

Let an observer A be placed on top of the Caucasus (fig. 6.1). His horizon will be bounded by conical surface ABE. An observer placed at I, at the base of the mountain, will have the circle DH as horizon. If the earth were at L, an observer on its surface would have the circle BE as horizon.[57] Thus, Galileo asserts, two observers at A and L would see the same portion of spherical space. Now, Galileo continues, the question is: How high would a mountain have to be in order to allow an observer on its summit to see the same portion of spherical space as that which would be visible to an observer placed at L? By making use of the Euclidean theory of proportion (and introducing an approximation concerning parallel lines), Galileo elegantly proves that the triangles shown in the figure (AOC, OCB) are similar. Subsequently, by adopting the standard astronomical values of his day for all the distances involved, he demonstrates by means of simple arithmetical calculations that the required height is slightly more than one mile. This means that placing the earth on the circle of the Sun causes a difference in the perception of the surrounding spherical space no greater than the difference which would be perceived on the summit of a one-mile-high mountain.[58]

Thus, in the final analysis, Aristotle's point on the illumination of the Caucasus, which is the basis of Mazzoni's objection, is totally irrelevant. The diagrams and the mathematical methodology proposed by Galileo are indexically dense. They allude both to Mazzoni's illustration of the Platonic passage from knowledge of shadows to knowledge of bodies and to Mazzoni's discussion and computations related to Aristotle's argument on the illumination of the Caucasus. One of Mazzoni's figures accompanying his demonstration of the passage from knowledge of shadows to knowledge of material bodies is like this. Both the sun and its circular trajectory on the celestial vault are clearly indicated, as well as the straight line along which light propagates. The shadow of the ornamentally delineated column is discernible on the ground. A sextant represents a tool of practical mathematics and observational astronomy.[59] Accordingly, the structure of Mazzoni's

57 Not any observer on the surface of the earth would have the circle BE as horizon, but Galileo seems to imagine the earth translated into L with the observer placed at the bottom of the mountain, namely, at I.

58 Galilei 1890–1909, II, pp. 200–202. It must be noted that Galileo's proof is further complicated by the fact that he needs to introduce a simplification, in that the two triangles can be proven to be perfectly similar only if lines AB and CB are thought of as being parallel to each other. Galileo argues that the difference in parallelism is negligible because of the enormous extension of the Copernican cosmos.

59 Mazzoni 1597, p. 182.

proof is based on practical arithmetic, and Mazzoni resorts many a time in his deduction to what he calls rule of the fourth proportional (sometimes also referred to in the contemporary literature as the rule of three, or golden rule), which is based on conception of ratios and proportions typical of practical arithmetic.[60] At another point of his book, in order to defend Aristotle from the criticisms of some mathematicians, Mazzoni further illustrates the argument on the illumination of the Caucasus with a diagram nearly identical to those forming the visual backbone of Galileo's counter-argument.[61]

I would suggest that visual and computational forms of indexicality are present in Galileo's hermetic performance.[62] Galileo's is taking his compositional cues (indexical presuppositions) both from Mazzoni's line of reasoning and from its visual and computational devices. But at the same time he is creative, in a very specific and, I would say, hermetic sense. "Creativity, in indexical terms, rests in introducing novelty in the form of a new topic or subtopic, or shifting the role relationships of the participants to the speaker's advantage, while satisfying the indexically presupposing constraints of the emergent".[63] The contents of Galileo's performance (the emergent) conforms to these aspects of indexical creativity. Galileo acts as though he and Mazzoni were once again taking part in the friendly Platonic conversations, or rather *ragionamenti*, he mentioned at the beginning of the letter. While he may not be introducing an entirely new topic (although the idea of the horizon as a conic surface can be construed as a new subtopic), he is certainly shifting the Platonic role relationship master-student to his advantage by taking the lead in mathematical argumentation, and by powerfully constraining his former

60 Mazzoni 1597, pp. 181 ff.

61 Mazzoni 1597, p. 197.

62 It is interesting to note Mazzoni's and Galileo's different uses of the first letter of the alphabet 'A'. Mazzoni uses 'A' for the centre of the Earth, while Galileo uses it for the summit of the mountain. Mazzoni's geocentrism is implicit in his unconscious strategy of lettering. In general, one might go as far as to suggest that many diagrams might be viewed as macro-indices living within the space created by the text. They are neither symbols nor icons and need to be accompanied by verbal descriptions in order for their elements to be 'located', that is, identified as tokens which refer to the objects already brought into play within a text. Diagrams do not speak for themselves.

63 Sawyer 1996, p. 280. Cf. also Sawyer 1992.

master's future response.[64] The emergent is a multi-layered structure indexically entailing, and at the same time satisfying the socio-cultural indexical presuppositions posed both by Mazzoni the former Platonic tutor during Galileo's juvenile years at Pisa and by Mazzoni the recent proponent of a four-topic theme on: (1) the Platonic passage from knowledge of shadows to knowledge of bodies; (2) mathematics as the form of knowledge that allows the philosopher to progress from shadows to bodies; (3) Aristotle and the Caucasus; (4) the Copernican system.

In a nutshell, in Galileo's performance, the emergent produces the net result of challenging Mazzoni's Platonism on the role of the mathematical natural philosopher. If Galileo is able to elaborate on mathematics to the extent that he can turn the tables on the Platonist and bring an implicit charge against his vision of mathematics as ancillary to dialectics, then the latter's epistemological hierarchy is automatically called into question. Thus, in his future response, Mazzoni is expected to take into due account this indexical entailment concerning Platonic epistemology. Let us now turn to the question of the disappearance of Galileo's counter-argument.

Antonio Favaro decided not to publish the Mazzoni letter in the volumes collecting Galileo's correspondence, but to place it in volume II, amongst Galileo's works, according to a chronological order.[65] Yet, in my view, the historian should resist the temptation to synthesize the textuality elements present in the Mazzoni letter into a Copernican text, and look at the letter as being first and foremost the vehicle of Galileo's mathematical performance.[66] Then the demise of the counter-argument will begin to appear less baffling.

---

64 By accepting Aristotle's authority on the pattern of illumination of the Caucasus, Galileo sets himself apart from the mathematicians targeted by Mazzoni in his defense of Aristotle. The direction towards which Galileo is steering the discussion neutralizes those concerns and focuses the performance on the significance of the question of the illumination of the Caucasus for cosmology.

65 In Favaro's view, the Mazzoni letter has only the form of the epistolary document, the substance of it justifies its being placed amongst Galileo's works (Galilei 1890–1909, II, p. 195).

66 This conclusion on the difficulty of establishing what exactly is the *text* contained in the Mazzoni letter can also be seen in the more general context of the question of the status of text in literary, anthropological, and linguistic studies of discourse. W. F. Hanks has recognized five dimensions present in the relevant literature. According to him, the status of text question breaks down into various problems coming under the following five headings, "the boundaries of text, the 'centering' of text within some interpretive matrix, text as performance, reception of text, and the construction of synchrony"(Hanks 1989, p. 103). My discussion has stressed the performance-like quality of Galileo's reaction

A fleeting performance was not enough to entextualize those elements into a more stable product such as a literary text. Too much of that performance depended on improvisational creativity and on the contingency of the circumstances to allow the accomplishment of the process of entextualization. "Unlike compositional forms which result in text artifacts, such as musical scores, written prose novels, paintings, or room interiors, all performances are creative processes which generate ephemeral products, essentially bound in and by the durational interval of the performance".[67] If we evaluate the episode staged through the medium of the Mazzoni letter along Sawyer's structural dimensions of improvisation (cf. Table 6.1), we observe that most of them are to a degree recognizable. Dimensions 2,3). We have seen how indexicality operates at both the socio-linguistic level and the philosophical-mathematical level. Dimension 4). Galileo refers to the contents of his letter to Mazzoni as proposing another friendly Platonic discussion [ragionamento], even though he is not actually talking to Mazzoni in a real oral setting. Dimension 6). Audience involvement is another important point, but on this question we can hardly go beyond speculation. However, not only did Galileo explicitly invite Mazzoni to answer his counter-argument, but we know that Galileo circulated the letter amongst his most intimate friends and that the letter would have been exposed to the curiosity of philosophers and literati at Padua.[68] Dimensions 7, 8). As we have seen, more improvisational genres are more receptive to novelty, and innovation is unlikely to last long. Accordingly, Galileo's counter-argument was receptive to the Copernican novelty, even though the counter-argument later disappeared.

Before concluding let me briefly comment on another salient aspect of this episode in connection with its interpretation as a performance targeted at Mazzoni's Platonic teaching. Galileo's response to Mazzoni's indexical entailments was immediate; in addition, if we agree with Sawyer's analysis of

to Mazzoni's book communicated by the Mazzoni letter. Hanks's analysis suggests that it would be rewarding further to explore the Mazzoni letter in relation to those issues.

67 Sawyer 1996, p. 281.

68 Favaro quotes a piece of information from Paolo Gualdo's biography of Vincenzo Pinelli (the leader of a famous literary group in Padua), both of whom we know were amongst Galileo's most intimate friends, according to which, Galileo's letter had been printed and was present in Pinelli's then renowned library. However, Favaro is absolutely convinced that Gualdo went wrong and that in Pinelli's library the letter was present in manuscript form only. Be that as it may, this suggests that the letter had voluntarily been made available to the wider public of scholars who had access to Pinelli's library. Cf. the details in Galilei 1890–1909, II, p. 195, and Favaro 1966, I, pp. 121 ff.

improvisational creativity (quoted above), we have to admit that some elements of it may have been not consciously goal-directed. But to what extent can we accept that Galileo's response was not consciously goal-directed? There is a sense in which, I believe, we can accept this surprising consequence, a sense which emerges if we once again look at the contents of the letter resisting the temptation to focus on its pro-Copernicus answer. My suggestion is along the following lines. In an intriguing study on improvisation and creativity in the arts, Sawyer discussed the relation *problem-solving* vs. *problem-finding*.[69] Whereas, for example, in a traditional theatre performance, the actors start with a script and the performance may be seen as an instance of problem-solving, in improv theatre the actors "have to create everything; the dramatic elements emerge from the dialogue, in a problem-finding process that is collaborative and emergent".[70] Accordingly we might consider Galileo's performance and the emergence of its creative content as a process of problem-finding, rather than of problem-solving. Remember how Galileo's counter-argument started out: *How high would a mountain have to be in order to allow an observer on its summit to see the same portion of spherical space as that which would be visible to an observer placed at L?* Finding the height (problem-solving) may have been an easy task given Galileo's mathematical skills, but he tells Mazzoni very clearly that formulating the question (problem-finding) cost him a lengthy discourse (*un lunghissimo discorso*).[71] Let us note that the question initiating the whole creative process by no means presupposes a Copernican goal. At the end of the calculation process Galileo might well have found that the mountain's height turned out to be one meter, one mile, one thousand miles, or even greater than the distance earth-sun.

---

69 Sawyer 2000, pp. 153–154. For Sawyer, "the improvisational style, called problem-finding by creativity researchers, is used by most successful painters [...]. A 'problem-finding' painter is constantly searching for her or his visual problem while painting – improvising a painting rather than executing one. In contrast, a problem-solving style involves starting with a relatively detailed plan for a composition and simply painting it" (*ibid.*, 153).

70 Sawyer 2000, p. 154.

71 Galilei 1890–1909, II, p. 199. Furthermore, the interpretation of the counter-argument as the result of a problem-finding process would be fully consistent with Galileo's not questioning Mazzoni's acceptance of Aristotle's authority on the pattern of illumination of the Caucasus. Indeed Galileo's counter-argument posed a newly found problem for Mazzoni. We can only regret that Mazzoni died one year after the publication of his book, probably without having the opportunity of answering Galileo and thus of entering into the dynamics of the performance.

In sum, I believe, there is a sense in which we can accept that surprising consequence. The dynamics of the problem-solving process is collaborative, not confrontational, and does not presuppose goal-directedness, at least not in an explicitly overt manner. That spur-of-the-moment flurry of creativity in favour of Copernicanism might not have been consciously goal-directed, but rather the serendipitous result of a Platonic exercise in the clearing of the mind. The piece of mathematical bravura disappeared, and for the next thirteen years Galileo was unable to come up with anything else to support Copernicus. This fact confirms Shea's intuition, i.e., that the Mazzoni episode might simply have raised Galileo's awareness of the potential of mathematics for cosmology.

Galileo's Maieutic Experiment

There are different perspectives from which the questions concerning maieutics and the theory of recollection raised by the 1632 *Dialogue* could be approached. Firstly, the *Dialogue* as a whole might be read as an implicit argument vaguely supporting the theory that the acquisition of knowledge is in essence a dialogical process whose function it might be to allow persons to recollect what they already know. Secondly, there are the single episodes in which the maieutic method is deployed in a more conscious and direct manner. I will concentrate on the second perspective, focusing in particular on a maieutic-mathematical episode. It centers around what Galileo refers to as Ptolemy's argument against the daily rotation of the Earth, an argument which is based on the idea that the circular motion of the earth would cause the extrusion of all bodies from its surface. In this episode, Galileo's mouthpiece Salviati questions the Aristotelian Simplicio and elicits answers from him in the Socratic manner. The episode has a final, surprising twist, in which Simplicio disowns all that he has conceded until that point while Salviati goes on to reject Ptolemy's argument by offering the paradoxical proof that, in spite of the rotational speed of the earth, if the bodies on its surface possess a vertical motion towards its centre, no matter how small, they will not be extruded. Within the episode Galileo encapsulates his most elaborate reflection on maieutics and the theory of recollection. It is the only point in the *Dialogue* at which Galileo has the Aristotelian Simplicio explicitly ask Salviati whether he endorses Plato's theory that our knowledge is a certain form of reminiscence (*nostrum scire sit quoddam reminisce*). Salviati refuses to answer Simplicio's question directly, claiming that

> [h]ow I feel about Plato's opinion I can indicate to you by means of words and also by deeds. In my previous arguments I have more than once explained myself with deeds. I shall pursue the same method in the matter at hand [i.e. Ptolemy's argument], which may then serve as an example, making it easier for you to comprehend my ideas about the acquisition of knowledge…[72]

Referring to this episode Koyré claimed that Galileo is utterly serious and that the allusions to Plato and the mention of the Socratic maieutics are "not superficial ornaments born from his [Galileo's] desire to conform to the literary mode inherited from the concern of Renaissance thought with Plato. […] they must be taken at their face value".[73] Shea has been more cautious about Salviati's answer, and has admitted that Galileo the philosopher was unwilling to avow explicitly the Socratic maieutics and the theory of recollection, and instead chose to illustrate his view in an artistic fashion.[74]

I suggest that the whole maieutic-mathematical episode is written in a comic style aimed at parodying maieutics and the theory of recollection. Galileo, in other words, indirectly tells us that without certain preconditions neither maieutics nor the theory of recollection are to be trusted in the search for knowledge. Above all, he tells us that mathematical argumentation is the most preferable way of securing knowledge of the natural world. To articulate this thesis, I will argue that in the maieutic-mathematical episode there are very specific textual characteristics which allow us to discover this fundamental parodic attitude. The most important of them is this. At the heart of the episode Salviati makes a remark to Sagredo concerning the 'certain form of reminiscence' (*quoddam reminisci*). Its strong ironic tone, one almost deriding Simplicio's sluggish groping for an answer, reveals Galileo's parodic attitude and the comic intent of the whole scene. This fact has so far escaped the notice of all commentators and translators because the irony in the remark has not been recognised. But before expanding on how we can discover the irony beyond Salviati's remark I must discuss how the complexity of ironic utterances has been approached in pragmatics. Irony has attracted the attention of scholars from many disciplines ever since Quintilian's classical definition. We are all familiar with Quintilian's analysis, according to which irony is simply

72 Galilei 1967, p. 191.

73 Koyré 1943, p. 427.

74 Shea 1972, p. 154.

saying something while meaning the opposite of it. Yet many ironic situations do not correspond to Quintilian's definition, in the sense that that definition does not always distinguish irony from non-irony. For example, a mother asks her children to clean their room. After a while she enters the still messy room saying to them 'I love children who keep their rooms clean'. The mother's utterance is literally true and does not mean the opposite of what it asserts. Yet irony is clearly how the mother conveys the message to her children.[75]

Akira Utsumi has furnished a very sophisticated and very general model of irony which seems to be able to capture most ironic situations. Utsumi's theory is based on the fundamental notion of 'ironic environment'. Ironic environment is a prerequisite for the speaker to be ironic. An ironic utterance achieves its effects by implicitly displaying the ironic environment by means of linguistic means. Let t0 and t1 be two instants that temporally precede the time when an utterance is given. The three following events/states define ironic environment.

1. The speaker has a certain expectation E at time t0.
2. The speaker's expectation E fails (i.e., E is incongruous with reality) at time t1.
3. The speaker has a negative emotional attitude (e.g., disappointment, anger, reproach, envy) toward the incongruity between what is expected and what actually is the case.[76]

When the discourse context includes these three events/states then the situation is said to be surrounded by ironic environment. In the above example of the mother entering her children's room the situation is surrounded by ironic environment. The ironist (the mother) has an expectation that the room is clean (a fact implied by her asking her children to clean the room), but the expectation has failed and we can reasonably assume that she is disappointed or even angry.[77] "The main claim of the implicit display theory is that verbal irony implicitly displays the fact that the situation is surrounded by ironic environment".[78] On the other hand, utterances are not ironic when they directly express at least one of the three components of ironic environment. Thus, for example, had the mother said '...this room is still in a mess!',

75 Utsumi 2000, p. 1779.
76 Utsumi 2000, p. 1783.
77 Utsumi 2000, pp. 1783–1784.
78 Utsumi 2000, p. 1784.

irony would have disappeared because the mother would have expressed her negative attitude directly. It is essential to recognize that Utsumi's theory suggests that in many communicative interactions irony can be present in various degrees. Therefore, the three elements constituting ironic environment should be seen as three dimensions defining an ironic space, so that a situation would be located within the ironic space according to the values of the three coordinates.[79] There would therefore be points maximally opposed which correspond to an ideally ironic situation and to a non-ironic situation. We need not dwell on the details and sophistication of the theory since our purposes are very limited, consisting simply in bringing to light the ironic environment that Salviati's remark displays. We can turn to the maieutic-mathematical episode. Let us quote the excerpt leading to Salviati's remark.

> Sal. Then let us proceed. Simplicio, tell me what motion is made by that little rock, tight in the notch of the stick, when the boy moves it so as to cast it a long way?
> Simp. The motion of the stone while it is in the notch is circular; that is, it travels along the arc of a circle whose fixed centre is the shoulder knot and whose radius is the stick and the arm.
> Salv. And when the stone escapes from the stick, what is its motion? Does it continue to follow its previous circle, or does it go along some other line?
> Simp. It certainly does not go on moving around, for then it would not fly away from the thrower's shoulder, and we should not see it go extremely far.
> Salv. Well, then, what is its motion?
> Simp. Let me think a moment here, for I have not formed a picture of it in my mind.
> Salv. Listen to that, Sagredo; here is the *quoddam reminisci* in action, sure enough. Well, Simplicio, you are thinking a long time.[80]

We now need Salviati's remark in the original. "Signor Sagredo, udite all' orecchio: ecco il *quoddam reminisci* in campagna, bene inteso. Voi ci pensate molto, Signor Simplicio!".[81] Here translators and commentators have missed the punch line, i.e. the metaphor 'in campagna' used by Galileo. The 'in action' is far too weak to capture the violence of the punch, so to speak, because 'in campagna' is taken from the semantic sphere of large-scale military campaigns, of armies fighting in the open field.[82] Salviati's punch line displays

79 Utsumi goes on to define a degree of ironicalness as the sum of the three values which the three coordinates can assume (cf. the details in Utsumi 2000, pp. 1787 ff.).

80 Galilei 1967, p. 191.

81 Galilei 1890–1909, VII, pp. 217–218.

82 The *Great Dictionary of the Italian Language* lists a number of expressions from sixteenth- and seventeenth-century literature in which the military meaning of 'in campagna' is well documented. Cf. *campagna*, sub vocem, in *Grande Dizionario della Lingua Italiana*.

the ironic environment created by Galileo in the context of the passage. Salviati's has an expectation that Simplicio will duly come up with answers in a reasonable time, but this expectation fails because Simplicio is sluggish. And the military metaphor used by Salviati in whispering to Sagredo's ear almost makes a mockery of the *quoddam reminisci*, disproportionately picturing it as an open field battle within Simplicio's mind. Salviati's metaphor unambiguously suggests the incongruity between his expectation and the reality of the *quoddam reminisci* in action. Basically, all the conditions required by the theory of irony based on the notion of indirect display of ironic environment are satisfied by the maieutic-mathematical episode. But we can go further and explore how in this episode Galileo is really in the business of parodying the Socratic method and the theory of recollection.

What is essential in order to form a parody? According to Rossen-Knill's and Henry's model of verbal parody, in every occurrence of verbal parody, "the speaker conventionally makes use of four essential acts: (1) the intentional verbal re-presentation of the object of parody, (2) the flaunting of the verbal re-presentation, (3) the critical act, and (4) the comic act".[83] Let us see how this characterization applies to our case. Galileo has intentionally re-presented the object of parody (the Socratic method and the theory of recollection) and flaunted his verbal re-presentation by having Simplicio acquiesce to Salviati's questioning in earnest before the latter's ironic utterance to Sagredo. Moreover, "[t]he critical act ridicules the object of parody".[84] As we have seen, the military metaphor chosen by Galileo doubtless ridicules Simplicio's effort. Finally, the comic act is the result of the integrated use of the three components already mentioned in order to produce a comic effect. But how must the first three acts be constrained to create a comic act? The "acts must be manipulated to create some incongruity".[85] Galileo exploits exactly this strategy. After a lot of questioning by Salviati, Simplicio, finally on the very verge of re-discovering the solution to the problem, suddenly has second thoughts exclaiming: "But how stupid of me! [...] And I allowed myself, simple-mindedly enough, to be convinced that stones would not be extruded by the whirling of the Earth! I take it back, then, and declare that if the earth did

83 Rossen-Knill and Henry 1997, 723. It should be pointed out that this model, precisely as Utsumi's model of irony, suggest that the four components of the parodic act may not be present together with the same level of explicitness. In our case, we shall see that they are all fairly well recognizable in the maieutic-mathematical episode.

84 Rossen-Knill and Henry 1997, p. 735.

85 Rossen-Knill and Henry 1997, p. 736.

move, then stones, elephants, towers, and cities would necessarily fly toward the heavens; and since that does not happen, I say that the earth does not move".[86] How incongruous, to say the least, that Simplicio cowardly gives up after so much laborious recollecting! Since Simplicio's inglorious retreat has cleared the battlefield, Salviati can clinch the argument by presenting a mathematical proof, a paradoxical one to be sure, according to which no heavy body would be extruded from the surface of a revolving earth providing that it had some vertical motion towards the centre, no matter how swift the earth rotated around its axis and how small the body's vertical motion were. We need not analyze the proof in detail. What matters most is simply the fact that not only has the maieutic experiment failed so miserably, but the mathematical proof necessary to disprove Ptolemy's argument is inaccessible to Simplicio, and, Galileo seems to imply, to the maieutic method and its attendant theory of reminiscence.

In sum, Galileo raises doubts about the validity of the Socratic method and the theory of recollection. Moreover, he seems to suggest that mathematics holds the key to correctly reasoning about the natural world, and that that key is beyond the grasp of both Simplicio the Aristotelian philosopher and Simplicio the dabbler in the Socratic method. This conclusion, however, should not be accepted without certain caveats. One might object that it is precisely Simplicio's having been adulterated in the Peripatetic school that prevents him from bringing to fruition his engagement in the Socratic method. We might re-formulate our previous conclusion asserting that Galileo implicitly posed a precondition for the proper functioning of the Socratic method, namely, the freedom from the type of indoctrination that he despised in the Aristotelian Simplicio. However, why, then, not have Sagredo, the curious layman not educated in the philosophical schools, impersonate Meno's slave and happily bring to a close the maieutic-mathematical exercise? Why make it fail so grotesquely? Although we cannot answer this question with certainty, I believe that it is fair to suggest that, by the time he published the *Dialogue*, Galileo had serious misgivings about the validity of the Socratic method and the Platonic theory of recollection.

---

86 Galilei 1967, p. 196.

The Comedy of the Heaven

The *Dialogo* is a great book about 'performing' the heaven. The archetype of the heaven unfolds in the theatrical discussions among Simplicio, Sagredo, and Salviati. Galileo's hermetic imagination represents itself as receiving the revelation of the new astronomy. But the language is replete with cues that point to a comedic style. Translations of Galileo's Italian prose have struggled to retain the nuances that announce Galileo's innuendos, parodies, and ironies. His grotesqueries suggest a background in the Italian *commedia dell' arte*. Plato's theory of the *quoddam reminisci* was mocked in a representational style that is strongly reminiscent of the exaggerated feats boasted by certain characters populating the stages of the *commedia dell' arte*. The *Dialogo* projects possibilities of understanding the heaven that might have been further elaborated in oral settings among friends, but which Galileo could not develop in writing for obvious reasons. It is like the tip of an iceberg that we cannot see.

The extrusion effect of the diurnal rotation of the Earth is presented by Galileo as one of the strongest objections against Copernicus. It was taken to be a powerful argument showing that the diurnal rotation of the earth is impossible. "Now there remains the objection based upon the experience of seeing that the speed of a whirling has a property of extruding and discarding material adhering to the revolving frame. For that reason it has appeared to many, including Ptolemy, that if the Earth turned upon itself with great speed, rocks and animals would necessarily be thrown toward the stars, and buildings could not be attached to their foundations with cement so strong that they too would not suffer similar ruin".[87] The three interlocutors are agreed that any circular motion, like that of a sling, or a wheel, has a faculty of extruding objects placed on the circumference, and that the direction of the object's motion upon leaving the extruding device is the tangent to the circumference at the point of separation. Further, they agree that the motion after separation will be uniform and that if the circular motion is fast enough extrusion in slings and wheels will at some point occur. But all heavy bodies on the earth's surface, unlike objects attached to slings, have a natural tendency downwards. The three interlocutors have no doubt on this either. Then Galileo issues his challenge to Simplicio.

Galileo wishes to prove that *no matter how small that downward tendency, and no matter how fast the diurnal rotation of the Earth, all heavy objects will remain firmly attached to the earth's surface*. Two objections are subsequently

87 Galilei 1967, p. 188.

raised by Sagredo. I here summarize and paraphrase Galileo's text. Imagine a body a few instants along the tangent after leaving the rotating earth. Immediately upon leaving the earth's surface it will start descending toward the centre of the earth in naturally accelerated fall. The first objection is as follows. The downward tendency, in terms of degrees of speed of fall, decreases ad infinitum as the body is thought of as approaching backwards the point of separation. Galileo has in fact already introduced in the *Dialogo* the law of falling bodies, and the idea of the uniform increase of a falling body's degree of speed with time. Second objection, how about the weight of the object? Going by Aristotelian physics the lighter the body the less fast it will fall. Thus, Sagredo concludes, by combining the two effects one has good reason to doubt that at least some objects will be able to escape the grip of the earth and eventually be extruded.[88] However, extrusion does not occur, even under the circumstances hinted at by Sagredo. This demonstrative feat by Galileo is extraordinary. The fact is that Salviati-Galileo is deeply intrigued by Galileo-Sagredo's objection but wants to show that extrusion will never occur. Since the curvilinear angle between the circular profile of the earth and the tangent at any point is infinitely narrower and more acute than any rectilinear angle, then the downward tendency of falling bodies, no matter how small, as long as it operates, will always be more than enough for them to cover the infinitesimal distance between the tangent and the surface of the earth. The argument appears like an impossible proof mixing up geometry and physics. It has to work no matter how small the tendency downwards, and no matter how fast the rotation of the earth. It magically overcomes the intuition about slings and spinning devices. This undertone of achieving the impossible makes the whole discussion a literary feat in its own right. Galileo plays with the intuition that spinning will not extrude objects. Boasting and incredulity are the personae of the discussion. There is diffused in the *Dialogo* an inflation of the epistemic ego. It feels omnipotent.

It is the same sensation that we find in the impossible stories told by Capitano Spavento, the character played by the famous *comico dell' arte*, Francesco Andreini.[89] Unlike Galileo's ego that plays with feats of knowledge, Capitano Spavento boasts about his feats in heaven. Andreini calls his feats 'bravure' (something like, *bravados*). His *canovacci* tell impossible stories about conquering the pagan heaven, often for no apparent reason. His heroes are the gods of the Greek pantheon. In one instance, he calls his thought to assault

88 Galilei 1967, pp. 188 ff.

89 Andreini 1615.

the heaven 'noble' and 'belligerent'.[90] But he does not tell us why exactly. His projection towards the heaven makes us feel like touching a place in close proximity with the earth. His heaven is not distant from the earth. Distance and proximity are not geometrical, though. They are existential. His touchable heaven is part of the imagination as if it were present in the life-world that we all (back in the seventeenth century) naturally inhabit. He makes the heaven 'behave' by means of his fantastic 'bravure'. Galileo makes the earth behave. It must keep its trees and buildings and people and animals glued to its surface. Galileo's epistemic bravura demonstrates that it is possible to keep order on the surface of the rotating earth. Capitano Spavento climbs mount Olympus and begins to throw arrows to heaven like mad.[91] Seeing the damage that our Capitano is doing Jove organizes the resistance. But the Capitano jumps to heaven and makes Jove a prisoner. He scolds him. He wants to put an end to Jove's dominion over the affairs of humans. Jove will no longer carry the title of nature, or of fate, or of providence, which are all superstitions of the populace. The Capitano argues that he is going to take Jove's place in the cosmic order. But eventually Venus mediates between him and the defeated god, asking for his liberation. The Capitano has compassion, frees Jove, and with a big jump from heaven lands safely in Venice. There is no space here for contemplating all the glorious feats of the Capitano. His *canovacci*, in which fabulous narrative, social criticism and philosophy are fused, hint at a performance style that captures the imagination.

90 Andreini 1615, p. 4 recto.

91 It is hard to render in English the phantasmagoric style of the story and the immediacy of the simple language. One wonders how that would have been played on the stage. Here is an excerpt. "Io vedendo l' essercito della Celeste Corte tutto schierato per muovermi crudelissimo assalto, che feci? Tirai una balestrata, e diedi d'una piramide nel mostaccio a Giove. Giove subito prendendo l'istessa piramide, e trasformandola in fulmine ardentissimo, me la slanciò nella testa; io ripigliando il fulmine con la destra mano, di nuovo lo slanciai nel Cielo, e dietro ad esso spiccai un salto con tanta destrezza, ch'io giunsi, et arrivai tra le Celesti squadre; là dove gionto, posi mano a questa mia tagliente, e fulminante spada, posi in fuga tutto l'essercito, e quivi facendo mio prigioniero il Regnator de l'Etere, e legandolo al fuso dell' Hemispero, cominciai a soridarlo con acre rampogne, prorompendo queste iraconde parole…" (Andreini 1615, p. 5).

The heaven that Galileo had expanded with his marvelous discoveries, as he told Diodati, remained in existential proximity with the earth, and was still populated by odd though innocuous characters. This archetypal heaven was not an exteriority which unperturbed consciousness reflected. It was a space of the imagination. It was the theater of incredible performances that challenged the intellect. Galileo does not review arguments and counter-arguments in a scholastic style. He performs them before the readers, and above all he performs them in the heaven of his hermetic imagination.

· 7 ·

HOSPITALITY

Like Capitano Spavento I will now jump back from the seventeenth century to the twenty-first. I will dedicate my final meditation to an *experiment* in hermeticism that transformed my life-world. By way of concluding this book I will offer some reflections drawn from my teaching experience and pedagogy. The latter helped me understand elements of hermeticism that I had not been able to appreciate on the basis of a purely historical and philosophical approach to the hermetic literature, its reception, and its influence on the Western mind. Teaching hermeticism helped me realize that doctrinal contents are not the only perspective that matters. It is important to think of hermeticism as an experience, or an encounter, rather than merely as a set of theories about human beings, God, and the cosmos.

The Academic Setting

In 2003 I was hired as an assistant professor in the department of History and Philosophy of Science at the University of Pittsburgh. I was to replace Ted McGuire who had decided to retire. Someone was needed to teach regularly a course, called *Magic, Medicine and Science*, that Ted had originally designed and been teaching for three decades. Ted's course focused on the legacy of hermeticism in Western thought until Newton. Over the last few years, I have experimented with many different formulas for *Magic, Medicine and Science*, especially with the dramatic challenges posed by teaching large classes. The presupposition of the course fascinated me. It drew attention to esoteric commonalities which, although they have shaped the history of Western civilization, have tended to be downplayed or outright rejected in academia by the predominant positivist ideology of the last century. These undercurrents of thought and myth were never fully absorbed into the mainstream Greco-Christian intellectual structures that form the backbone of modern science. But an innovative way was needed to communicate this fascination to young students, to make it into stuff that would stand a chance to enhance their lives. For what other purposes does teaching have than the attempt to make young lives find their own mysterious pathways to knowledge, health, emotional awareness, and communal wellbeing? I had grown more and more dissatisfied with the notion of teaching as mere transmission of doctrines, a process of manipulation of consciousness, based on indoctrination and propaganda, that has transformed higher education institutions into branches of the multinational corporate state. Hence I started to think about a pedagogy that placed the learner at its center as the protagonist of a struggle for emancipation. At the same time I realized that my teaching would not improve if I was not willing to go through a process of radical self-criticism, which implied re-visioning the positivist ideology that informed the field of the history and philosophy of science. What gradually emerged over the last ten years or so from the collegial effort of the learners was the discovery of the ethnicity of reason and its revolutionary potential for social and educational reform. The learners involved in the experiment came to realize that the current system of higher education, superficially enthused with the notions of tolerance, non-discrimination, diversity, and integration, is in reality desperately trying to eradicate the ethnic background of reason that challenges the subservience of our educational institutions to worldwide economic interests. Exploitative capitalism dictates what we should learn in terms of contents, how we should classify and disqualify human beings according to a grading system that turns students into marketable products, who is and who is not allowed to show up in the classrooms, what we should think in terms of welfare, health science and ecology, and what we can hope to achieve as learners and what we should not dare to call into question. *Magic, Medicine and Science* became a gateway to understanding much more than science processes and their history. It became a pedagogy in its own right.

Asclepius in the Classroom

In 2015 *Magic, Medicine, and Science* was proposed to a class of thirty undergraduates as an open platform for student emancipation and self-exploration based on the most complex text of the hermetic literature, the *Asclepius*. We alleviated the anxiety of the grade by announcing that the destructive violence of the assessment process, required by our institution, was going to be minimized and based solely on the completion of a portfolio whose individual sections were not going to be graded. We would not interfere with the student's freedom to earn the highest grade. We acted in defiance of the branding of cattle that brutalizes our educational institutions. The structure of the class was simple. Every week was devoted to an evocative theme from the *Asclepius*. A brief lecture introduced the theme at the beginning of the first meeting of the week. Small group discussions of the possible ramifications of the theme followed under the stimulus of the participants. The second meeting of the week continued the dialogic process in multiple directions. We regularly proposed hand-outs as the basis for discussion and reflection on topics that might be evolved in view of the portfolio. The final choice of the media and topics for the portfolio was left to the students. Possibilities included essay writing, creative prose, poetry, music, visual arts, analysis of dreams, and mixed media. The class started with an effort to create the learning environment that we thought would be most conducive to freedom and growth. The first week was dedicated to the exchange of intellectual gifts intended as a means of strengthening friendship in and beyond the classroom. The exchange was the clearing in the forest that oriented the value system that was to frame the reception of hermeticism. Students in pairs were invited to exchange letters in which they needed to write personal notes of appreciation for each other, following a brief interview in which they had learned about their friend's personal history, ethnic background, aspirations and common and/or conflicting beliefs. Every week volunteers shared their work and reflections with the classroom. The twelve themes that were highlighted in the course and which formed the sections of the portfolios were: (1) The Corpus Hermeticum, (2) The one, (3) Archetypes, (4) Mind and body and the divine, (5) Human beings and the cosmos, (6) Philosophy, (7) The natural world, (8) Love, (9) Life and death, (10) Eternity and time, (11) Consciousness, (12) The senses. The entire class completed the required portfolios.

The most surprising element that emerges from the composition and structure of the portfolios is the autobiographical emphasis. When students are left free to engage their own potential for learning in unforeseen directions, they tend to turn inwards and towards their personal histories in search of clarity of thought. The doctrinal contents that are transmitted through studying a philosophical text, no matter how far removed from the actuality of the present (and the *Asclepius* is certainly in this category, at least from the viewpoint of undergraduate students), are evaluated and digested through a process of self-exploration, bewilderment, and the discovery of shared interests. Regardless of the lecturer's efforts to clarify the contents that are being evaluated, the response is always mediated by an engagement of the whole personality that is decisive in steering the appropriation of the material. Clarity of thought does not seem to be achieved only through clarity of exposition though the latter is certainly a desideratum. It is rather the outcome of a vastly more complex process of growth that must recuperate energies which are buried in personal history and family and ethnic traditions. This was most surprising to us. We noticed that the climate of the class changed over the course of the semester. As the students became more and more familiar with one another and with their instructors, the process of learning gained momentum. But it quickly became apparent that the students themselves were taking the lead. No matter how good or how awful my brief lectures and commentaries were, clarity of vision came through slowly but surely. It was a new experience of clarity that, I learned, was not conceptual, in the sense that it was not the result of reasoning from first principles.

I emphasize that there were no learning objectives explicitly stated at any time. These were left to be discovered through the personal and collegial processes of learning. We thought that one way to weaken the stifling pressure for *achieving*, the most virulent and destructive virus that has infected our educational systems, was to refuse to set objectives for anybody. Nobody was under pressure to achieve anything. Indeed, we made it clear that we wanted everybody to feel free to engage with the class, responsibly, in any manner they thought was going to be beneficial to them and to their classmates. The emphasis on achievement and contents that pollutes the learning process can be alleviated through freedom, self-discipline, and enjoyment.

Hospitality

In the context of *Magic, Medicine, and Science* the collegial learning experience brought me into close contact with the exigencies of ethnicity and hospitality. The authoritarian ideology of the classroom in which an authoritative figure is supposed to transmit contents to an audience appears to me no longer tenable. The experiment in hermeticism has brought home to me the richness of ethnic traditions and personal histories. These will have to form the basis for a new direction in education. But it is also the case that a more responsible direction in the historical and philosophical studies of science will also be found when the ethnicity of reason is recognized. With this recognition comes the requirement for hospitality. I owe my intuition that hospitality is essential to the equanimity of both the learning and the scholarly processes, in the classroom and beyond, to Edith Doron, who taught me that you become genuinely hospitable by being yourself an exile in a wasteland eager to welcome a guest.

The experiment in hermeticism convinced me that teaching a content that already preexists in your mind, as a fixed, sacred, and unquestionable given received from a venerable tradition, implies firstly that you will not be hospitable to the absolute strangeness of the other being, and secondly that the condition for the other to be accepted into your castle is their conversion to your 'true' religion. There is also the other mind that resides within yourself, and which must be admitted into the house of consciousness so as to learn how to deal with the ethical obligation to welcome absolute strangeness. The *Asclepius* was a good choice for this exercise. It is an alien voice from an alien past that resounds mysteriously. It is an encounter with a stranger that asks for hospitality. How can we welcome such a guest in the house of science and technology and under the auspices of modern industrial civilization? Yet, to our surprise, the stranger spoke to the participants in the class in unsuspected modes that resonated with their lives profoundly. It has become clear to us in hindsight that what we were actually attempting to do was an experiment in hospitality. This is one lesson that we learned. I think it is a good lesson. We would not have learned this if we had formally stated learning objectives. For some of us *Asclepius* has become a companion in our house, but we do not pretend to know why.

Back to Galileo and Eliot

> Though you have shelters and institutions,
> Precarious lodgings while the rent is paid,
> Subsiding basements where the rat breeds
> Or sanitary dwellings with numbered doors
> Or a house a little better than your neighbour's;
> When the Stranger says: 'What is the meaning of this city?
> Do you huddle close together because you love each other?'
> What will you answer? 'We all dwell together
> To make money from each other' or 'This is a community'?
> And the Stranger will depart and return to the desert.
> O my soul, be prepared for the coming of the Stranger,
> Be prepared for him who knows how to ask questions.[1]

Perhaps the reader will want to know how all of this relates to Galileo, the telescope, and hermeticism. It is a legitimate question. I do not know how to answer. I can only ask for hospitality in the reader's mind. I conclude with a poem and an interpretation that might illuminate the reader's question. Eliot's poem "speaks to the difference between reception of the other as an interested act, as part of the struggle for recognition for the sake of being-at-home and therefore as an act that constitutes one's sovereign ipseity on one side, and the encounter with the other for whom no common frontier exists and yet whose proximity creates the uncanny dispossession of home, the internal sense of an undoing. It speaks to the non-ground of community for which 'I' and 'you' are not individuals of a common concept. Among the poem's diversity of dwellings, a 'we' emerges to perform the fiction of a social bond. The Stranger departs and returns to the desert—there where hospitality is a matter of life or death—but his utterance remains to make a desert of those dwellings in the mind where sovereign Identity lives as master; when for an instant astonishment makes Identity a guest".[2]

In the classroom and in the scholarly act of welcoming the other 'for whom no common frontier exists', as Edith Doron phrases it, lies the possibility of a deeper understanding of the absolute obligation to the stranger. The hermeneutic of encounter brings together the scholar, the student, and the teacher

1 From T. S. Eliot, *Choruses from 'The Rock'* (Eliot 1936, p. 103).
2 Doron 2013, pp. 191–192.

like travelers, resting and telling stories around a camp-fire, before they continue their journey in the desert. Encountering Galileo in the company of *Asclepius* and of many young learners who like myself were dispossessed of home following strange pathways was part of a journey towards relinquishing the sovereign ipseity that preempts the duty of hospitality.

BIBLIOGRAPHY

Acerbi, Fabio. 2000. "Le fonti del mito Platonico di Galileo". *Physis* 37, pp. 359–392.

Alfani, Augusto. 1872. *Della vita e degli scritti di Orazio Ricasoli Rucellai: Studio critico*. Florence: Tipografia Barbèra.

Alighieri, Dante. 1987. *Commedia*. Edited by E. Pasquini and A. Quaglio. Milan: Garzanti.

Allen, Don Cameron. 1970. *Mysteriously meant: The rediscovery of pagan symbolism and allegorical interpretation in the Renaissance*. Baltimore: Johns Hopkins Press.

Andreini, Francesco. 1615. *Le bravure del Capitano Spavento*. Venice: Vincenzo Smoasco.

Anguillara, Giovanni Andrea. 1584. *Le matamorfosi di Ovidio, ridotte da Giovanni Andrea dell' Anguillara in ottava rima, con le annotazioni di* M. *Gioseppe Horologgi, et gli argomenti, et postille di* M. *Francesco Turchi*. Venice: Bernardo Giunti.

Arias, Francisco. 1602. *Profitto spirituale, nel quale si insegna a fare acquisto delle virtù, e progresso nello spirito*. Translated by Giulio Zanchini. 2 vols. Venice: Appresso Fioravante Prati.

Ariosto, Ludovico. 1858. *The Orlando furioso translated into English verse*. Translated by William Stewart Rose. 2 vols. London: Henry G. Bohn.

———. 1999. *Orlando furioso*. Edited by Cesare Segre. 2 vols. Milan: Arnoldo Mondadori. Retrieved at: http://www.liberliber.it/

Armijo, Carmen. 2001. "Un nuevo rol para las definiciones". In: J. Montesinos and C. Solís [Eds.] *Largo campo di filosofare. Eurosymposium Galileo 2001*. La Orotava, Tenerife: Fundación Canaria Orotava de Historia de la Ciencia, pp. 85–99.

Ars chemica...Studium Consilii coniugij de massa solis et lunæ. 1566. Strasbourg: Samuel Emmel.

Avilès, Alejandro García. 1997. "Alfonso X y el *Liber Razielis*: imágenes de la magia astral judía en el scriptorium alfonsí". *Bulletin of Hispanic Studies* 74, pp. 21–39.

Baillet, Adrien. 1691. *La vie de monsieur Des-Cartes*. Paris: Chez Daniel Horthemels.

Balduin, Christian Adolph. 1675. *Aurum superius & inferius aurae superioris & inferioris hermeticum*. Amsterdam: Apud Joannem Jansonium à Waesberge.

Bauman R., and Briggs, C. L. 1990. "Poetics and performance as critical perspectives on language and social life". *Annual Review of Anthropology* 19, pp. 59–88.

Bellarmino, Roberto. 1858. *Opera omnia. Disputationum Roberti Bellarmini...De controversiis Christiane fidei adversus huius temporis haereticos. Tomus quartus*. Naples: Giuseppe Giuliano.

Bellucci, Maria. 1971. "La filosofia naturale di Claudio Berigardo (con appendice di lettere inedite)". *Rivista critica di storia della filosofia* 36, pp. 363–411.

Bérigard, Claude. 1632. *Dubitationes in dialogum Galilaei Galilaei*. Florence: Ex Typographia Petri Nesti sub Signo Solis.

Berno, Francesca Romana. 2008. "Appunti sul latino di Galileo Galilei". *Atti e Memorie dell' Accademia Galileiana di Scienze, Lettere ed Arti, già dei Ricovrati e Patavina. Volume CXIX (2006–2007). Parte III: Memorie della Classe di Scienze Morali, Lettere ed Arti*, pp. 15–37.

Biagioli, Mario. 1993. *Galileo, courtier. The practice of science in the culture of absolutism*. Chicago and London: The University of Chicago Press.

Billot, Louis. 1914. *De Ecclesiae sacramentis. Commentarius in tertiam partem S. Thomae. Tomus prior. Complectens quaestiones de sacramentis in communi, de baptismo, confirmatione, et eucharistia*. Editio quinta aucta et emendata. Rome: Ex Typographia Pontificia in Instituto PII ix.

Bocchi, Achille. 1574. *Symbolicarum quaestionum, de vniuerso genere, quas serio ludebat, libri quinque*. Bologna: Apud Societatem Typographiae Bononiensis.

Brocchieri, Vittorio Beonio. 2012. "From 'peasant' tenants farmers to *Bourgeois-Gentilhommes*: Social mobility and family identity in Habsburg Lombardy". In: G. de Luca and G. Sabatini [Eds.] *Growing in the shadow of an empire. How Spanish colonialism affected economic development in Europe and in the world (XVIth and XVIIth cc.)*. Milan: Franco Angeli, pp. 163–182.

Bruno, Giordano. 2002. *Opere italiane*. 2 vols. General Editor: Nuccio Ordine. Critical texts by Giovanni Aquilecchia. Turin: UTET.

Cantimori, Delio. 1936. "Anabattismo e neoplatonismo in Italia nel XVI secolo". *Reale Accademia dei Lincei. Rendiconti della classe di scienze morali, storiche e filologiche*. Serie 6, volume 12, pp. 521–562.

———. 1992. *Eretici italiani del Cinquecento* e *Prospettive di storia ereticale del Cinquecento*. Edited by Adriano Prosperi. Turin: Einaudi, 1992 (*Eretici* 1st ed., Florence: Sansoni, 1939, and *Prospettive* 1st ed., Bari: Laterza, 1960).

Cantimori, Delio, & Feist, Elisabeth. 1937. *Per la storia degli eretici Italiani del secolo XVI in Europa*. Rome: Reale Accademia d'Italia.

Carnap, Rudolf. 1936. "Testability and meaning". *Philosophy of Science* 3, pp. 419–471.

———. 1937. "Testability and meaning (continued)". *Philosophy of Science* 4, pp. 1–40.

Cartari, Vincenzo. 1581. *Le imagini de i dei de gli antichi*. Lyon: Apresso Stefano Michele.

Castellio, Sebastian. 1981. *De arte dubitandi et confidendi, ignorandi et sciendi*, with an introduction and notes by Elisabeth Feist Hirsch. Leiden: Brill.

Cellio, Marco Antonio. 1680. *Il fosforo o' vero la pietra Bolognese preparata per rilucere fra l'ombre*. Rome: per il Vannacci.

Chiaramonti, Scipione. 1633. *Difesa di Scipione Chiaramonti da Cesena al suo antiticone, e Libro delle tre nuoue stelle dall'oppositioni dell'autore de' due massimi sistemi tolemaico, e copernicano*. Florence: Appreso il Landini.

Chlup, Radek. 2007. "The ritualization of language in the *Hermetica*". *Aries* 7, pp. 133–159.

Ciliberto, Michele. 1976. "Intellettuali e fascismo. Note su Delio Cantimori". *Studi Storici* 17, pp. 57–93.

Clericuzio, Antonio. 2000. *Elements, principles and corpuscles: A Study of atomism and chemistry in the seventeenth century*. Dordrecht: Kluwer Academic Publishers.

Corradino, Corrado. 1880. *Il secentismo e l' 'Adone' del Cavalier Marino*. Turin: F. Casanova.

Couliano, Ioan. 1987. *Eros and magic in the Renaissance*. Translated by M. Cook. Chicago and London: The University of Chicago Press.

Cremonini, Cesare. 1613. *Disputatio de coelo*. Venice: Apud Thomam Balionum.

———. 1627. *Apologia dictorum Aristotelis de origine, et principatu membrorum adversus Galenum*. Venice: Apud Hieronymum Piutum.

Croce, Benedetto. 1899. *I trattatisti italiani del 'concettismo' e Baltasar Gracian*. Naples: Stabilimento tipografico della R. Università.

———. 1902. *Estetica come scienza dell' espressione e linguistica generale*. Milan/ Palermo/ Naples: Remo Sandron.

———. 1912. *Le fonti della gnoseologia Vichiana. Memoria letta all' Accademia Pontaniana nella tornata del 10 Marzo 1912 dal socio Benedetto Croce*. Naples: R. Stabilimento tipografico Francesco Giannini e Figli.

Curione, Celio Secondo. 1544. *Araneus, seu de providentia dei, libellus vere aureus,cum aliis nonnullis eiusdem opusculis, lectu dignissimi, nunque primum in lucem editis*. Basel: Ex Officina Ioannis Oporini.

Darwin, Charles. 1871. *The descent of man, and selection in relation to sex*. 2 vols. London: John Murray.

Dionysius, the Areopagite. 1566. *Opera omnia quae extant*. Paris: Ex officina typographica Michaëlis Vascosani.

———. 1897. *The works*. Translated by John Parker. London: James Parker.

Doron, Edith. 2008. *'Of things and thresholds': Exilic education, hospitality and the practice of community in the children's museum*. Masters of Modern Thought Dissertation. Aberdeen: University of Aberdeen.

———. 2013. "Welcoming the stranger. From social inclusion to exilic education". In: Vincent V. J. van Gerven Oei, Adam S. Groves, and Nico Jenkina [Eds.] *Pedagogies of disaster*. New York and Tirana: Punctum Books and The Department of Eagles, pp. 191–203.

Drake, Stillman. 1973. "Velocity and Eudoxan proportion theory". *Physis* 15, pp. 49–64.

———. 1974. "Mathematics and discovery in Galileo's physics". *Historia mathematica* 1, pp. 129–150.

———. 1987. "Euclid Book V from Eudoxus to Dedekind". *Cahiers d' histoire et de philosophie des sciences*, n.s., 21, pp. 52–64.

———. 1999. *Essays on Galileo and the history and philosophy of science.* 3 vols. Edited by Noel M. Swerdlow, Trevor Harvey Levere. Toronto, Buffalo, London: University of Toronto Press.

Duranti, Alessandro. 1997. *Linguistic anthropology*. Cambridge: Cambridge University Press.

Eliot, T. S. 1936. *Collected poems, 1909–1935*. London: Faber & Faber.

Favaro, Antonio. 1885. "Documenti inediti per la storia dei manoscritti Galileiani nella Biblioteca Nazionale di Firenze". *Bullettino di bibliografia e di storia della scienze matematiche e fisiche* 18, pp. 1–112 and pp. 151–230.

———. 1887. *La libreria di Galileo Galilei*. Rome: Tipografia delle Scienze Matematiche e Fisiche.

———. 1888. *Per la Edizione Nazionale delle opere di Galileo Galilei sotto gli auspicii di S.M. il Re d' Italia.* Florence: Tipografia di G. Barbèra.

———. 1899. "Intorno alle opere scientifiche di Galileo Galilei nella Edizione Nazionale sotto gli auspicii di S.M. il Re d' Italia". *Atti del Reale Istituto Veneto di Scienze, Lettere ed Arti. Anno Accademico 1898–1899. Tomo LVIII- Parte Seconda*, pp. 129–204.

———. 1907. "Pensieri, sentenze e motti di Galileo Galilei raccolti dai discepoli e pubblicati da Antonio Favaro". *Rivista di fisica, matematica e scienze naturali* 86, pp. 97–109.

———. 1920. "Oppositori di Galileo Galilei. IV. Claudio Berigardo". *Atti del Reale Istituto Veneto di Scienze, Lettere ed Arti* 79, parte II, pp. 39–92.

———. 1949. *Pensieri, motti e sentenze tratti dalla Edizione Nazionale delle opere*. Florence: Barbèra.

———. 1966. *Galileo Galilei e lo studio di Padova*. 2 vols. Padova: Editrice Antenore [1st ed. 1883. Florence: Successori Le Monnier].

Feldhay, Rivka. 1995. *Galileo and the Church: Political inquisition or critical dialogue?* Cambridge: Cambridge University Press.

Ficino, Marsilio. 1561. *Opera & quae hactenus extitere, & quae in lucem nunc primum prodiere omnia, omnium artium & scientiarum, maiorumque facultatum multipharia cognitione refertissima, in duos tomos digesta, quorum seriem uersa pagella reperies*. 2 vols. Basel: Per Henricum Petri.

Finocchiaro, Maurice. 1989. *The Galileo affair: A Documentary history*. Berkeley and Los Angeles: The University of California Press.

Frajese, Attilio. 1964. *Galileo matematico*. Rome: Studium.

Gabriele da San Vincenzo. 1658. *Tractatus tres. I. De gratia. II. De iustificatione. III. De merito. Iuxta Catholicam, et profundam D. Augustini clarissimi Ecclesiae luminis, et D. Thomae Doctoris Angelici, doctrinam*. Rome: Ex typographia Haered. Francisci Mancini.

Galilei, Galileo. 1610. *Sidereus nuncius*. Venice: Apud Thomam Baglionum.

———. 1890–1909: *Le opere di Galileo Galilei*. Edizione Nazionale. Edited by Antonio Favaro. 20 vols. Florence: Barbèra.

———. 1967. *Dialogue concerning the two chief world systems*. Transated by Stillman Drake. Berkeley and Los Angeles: University of California Press (1st ed. 1957).

———. 1992. *Sidereus nuncius. Le messager celeste*. Edited and translated by Isabelle Pantin. Paris: Le Belles Lettres.

———. 2000. *Lettera a Cristina di Lorena*. Edited by Franco Motta. Introduction by Mauro Pesce. Genova: Marietti.

———. 2009. *Galileo's Sidereus Nuncius or a Sidereal Message*. Translated from the Latin by William R. Shea. Introduction and notes by William R. Shea and Tiziana Bascelli. Sagamore Beach, MA: Science History Publications.

Galois, Évariste. 1962. *Écrits et mémoires mathématiques d' Évariste Galois*. Edited by R. Bourgne et J.-P. Azra. Paris: Gauthier-Villars.

Garin, Eugenio. 1957. "Chi legga di A. Koyré". *Giornale critico della filosofia italiana* 11, pp. 406–408.

———. 1978. *Storia della filosofia italiana*. 3 vols. 3rd ed. Turin: Einaudi.

Gassendi, Pierre. 1630. *Epistolica exercitatio*. Paris: Apud Sebastianum Cramoisy.

Giusti, Enrico. 1981. "Aspetti matematici della cinematica Galileiana". *Bollettino di storia delle scienze matematiche* 1, pp. 3–42.

———. 1986. "Ricerche galileiane: il trattato 'De motu equabili' come modello della teoria delle proporzioni". *Bollettino di storia delle scienze matematiche* 6, pp. 89–108.

———. 1990. "Galilei e le leggi del moto". In: G. Galilei, *Dicorsi e dimostrazioni matematiche intorno a due nuove scienze attinenti alla meccanica ed i movimenti locali*. Edited by Enrico Giusti. Turin: Einaudi, pp. ix–lx.

———. 1992. "La teoria galileiana delle proporzioni". In: Lino Conti [Ed.] *La matematizzazione dell' universo. Momenti della cultura matematica tra '500 e '600*. Perugia: Edizioni Porziuncola, pp. 207–222.

———. 1993. *Euclides reformatus. La teoria delle proporzioni nella scuola galileiana*. Turin: Bollati Boringhieri.

———. 1994. "Il filosofo geometra. Matematica e filosofia naturale in Galileo". *Nuncius* 9, pp. 485–498.

———. 1995. "Il ruolo della matematica nella meccanica di Galileo". In: Alberto Tenenti et al., *Galileo Galilei e la cultura veneziana*. Venice: Istituto Veneto di Scienze Lettere e Arti, pp. 321–338.

Grassi, Ernesto. 1983. *Heidegger and the question of Renaissance humanism*. Edited by Mario A. Di Cesare. Binghamton, NY: Center for Medieval and Early Renaissance Studies State University of New York at Binghamton.

———. 1988. *Renaissance humanism: studies in philosophy and poetics*. Binghamton, NY: Medieval & Renaissance Texts & Studies (translated by Walter F. Veit, from the German *Einführung in philosophische Probleme des Humanismus*. Darmstadt: Wissenschaftliche Buchgesellschaft, 1986).

———. 1994. *The primordial metaphor*. Translated by L. Pietropaolo and M. Sarci. Binghamton, NY: Center for Medieval and Early Renaissance Studies State University of New York at Binghamton.

Hall, Crystal. 2013. *Galileo's reading*. Cambridge: Cambridge University Press.

Hanks, W. F. 1989. "Text and textuality". *Annual Review of Anthropolgy* 18, pp. 95–127.

Heilbron, John. 2010. *Galileo*. Oxford and New York: Oxford University Press.

Husserl, Edmund. 1970. *The crisis of European sciences and transcendental phenomenology. An introduction to phenomenological philosophy*. Translated by David Carr. Evanston: Northwestern University Press [original edition *Die Krisis der europäischen Wissenschaften und die transzendentale Phänomenologie: Eine Einleitung in die Phänomenologische Philosophie*. Edited by Walter Biemel. The Hague: Martinus Nijhoff, 1954].

Jardine, Nicholas. 1991. "Demonstration, dialectic, and rhetoric in Galileo's Dialogue". In: D. R. Kelley and R. H. Popkin [Eds.] *The shapes of knowledge from the Renaissance to the Enlightenment*. Dordrecht/ Boston/ London: Kluwer Academic Publishers, pp. 101–121.

Keller, Vera. 2014. "Hermetic Atomism: Christian Adolph Balduin (1632–1682), Aurum Aurae, and the 1674 Phosphor". *Ambix* 61, pp. 366–384.

Kepler, Johannes. 1937–2009. *Gesammelte Werke. Herausgegeben im Auftrag der deutschen Forschungsgemeinschaft und der Bayerischen Academie der Wissenschaften. Unter der Leitung von Walther von Dick und Max Caspar*. 25 vols. published as of 2009. München: C. H. Beck'sche Verlagsbuchhandlung.

Kircher, Athanasius. 1674. *Turris Babel, sive Archontologia*. Amsterdam: Ex officina Janssonio-Waesbergiana.

Koyré, Alexandre. 1943. "Galileo and Plato". *Journal of the History of Ideas* 4, pp. 400–428.

———. 1978. *Galileo studies*. Translated from the French by John Mepham. Hassocks, Sussex: The Harvester Press Limited, 1978 (1st ed. in French, 3 vols., Paris: Hermann, 1939).

Lagalla, Giulio Cesare. 1612. *De phoenomenis in orbe lunae novi telescopii usu a D. Gallileo Gallileo nunc iterum suscitatis physica disputatio…necnon de luce, et lumine altera disputatio*. Venice: Apud Thomam Balionum.

Liceti, Fortunio. 1636. *Ulysses apud Circen, siue De quadruplici transformatione. Deque varie transformatis hominibus dialogus ethico-physicus*. Udine: ex typographia Nicolai Schiratti.

———. 1640. *Litheosphorus*. Udine: ex typographia Nicolai Schiratti.

———. 1642. *De lunae subobscura luce prope coniunctiones et in eclipsibus observata*. Udine: Typis Nicolai Schiratti.

———. 1653. *Hieroglyphica, siue antiqua schemata gemmarvm anularivm, qvaesita moralia, politica, historica, medica, philosophica, & sublimiora*. Padua: Typis Sebastiani Sardi.

Lippomano, Luigi. 1553. *Confirmatione et stabilimento di tutti li dogmi catholici, con la subversione di tutti i fondamenti, motivi e ragioni delli moderni heretici*. Venice: Nella Contra de Sanata Maria Formosa, al segno de la Speranza.

MacPherson of Cluny, Sir William. 1999. *The Stephen Lawrence Inquiry*. Advised by Tom Cook, The Right Revered Dr John Sentamu, Dr Richard Stone. Retrievable at: https://www.gov.uk/government/publications/the-stephen-lawrence-inquiry

Marzio, Galeotto. 1517. *De homine libri duo*. Basel: Apud Ioannem Frobenium.

———. 1552. *De doctrina promiscua*. Lyon: Apud Ioannem Tornaesium.

Mazzoni, Jacopo. 1597. *In universam Platonis, et Aristotelis philosophiam praeludia, sive de comparatione Platonis, et Aristotelis*. Venice: Apud Ioannem Guerilium.

McGuire, James E. 1977. "Neoplatonism and active principles: Newton and the Corpus Hermeticum". In: R. S. Westman and J. E. McGuire, *Hermeticism and the Scientific Revolution. Papers read at a Clark Library Seminar, March 9, 1974*. Los Angeles: University of California, William Andrews Clark Memorial Library, pp. 95–142.

Mentzel, Christian. 1675. *Lapis Bononiensis in obscuro lucens, collatus cum Phosphoro hermetico*. Bilefeldiae: Typis Iusti Trenckeneri.

Miggiano, Gabriella. 2008. "Marzio, Galeotto". In: *Dizionario Biografico degli Italiani*———Volume 71. Retrievable at: http://www.treccani.it/enciclopedia/galeotto-marzio_%28Dizionario-Biografico%29/

Monconys, Balthasar de. 1665. *Iournal des voyages des monsieur De Monconys*. Publié par le Sieur de Liergues son Fils. Lyon: Chez Horace Boissat, et George Remes.

Muratori, Lodovico Antonio. 1730. *Della perfetta poesia italiana. Tomo Primo*. Venice: Appresso Sebastiano Coleti.

Nettl, Bruno. 1974. "Thoughts on improvisation: A Comparative approach". *The musical quarterly* 60, pp. 1–19.

Nock, A. D., and Festugière, A.-J. 1945/1954. *Corpus Hermeticum*. 4 vols. Paris: Société d' Edition Les Belles Lettres.

Ongaro, Giuseppe. 2005. "Fortunio Liceti". In: *Dizionario Biografico degli Italiani. Rome: Istituto dell'Enciclopedia Italiana*. Retrievable at: http://www.treccani.it/enciclopedia/fortunio-liceti_%28Dizionario-Biografico%29/

Palladino. F. 1991. "La teoria delle proporzioni nel Seicento". *Nuncius* 6, pp. 33–81.

Palmer, G. B., and Jankowiak, W. R. 1996. "Performance and imagination: Toward an anthropology of the spectacular and the mundane". *Cultural Anthropology* 11, pp. 225–258.

Palmieri, Paolo. 1998. "Re-examining Galileo's Theory of Tides". *Archive for History of Exact Sciences* 53, pp. 223.–375.

———. 2001. "The obscurity of the equimultiples. Clavius' and Galileo's foundational studies of Euclid's theory of proportions". *Archive for history of exact sciences* 55, pp. 555–597.

———. 2003. "Mental models in Galileo's early mathematization of nature". *Studies in History and Philosophy of Science* 34, pp. 229–264.

———. 2007. "Science and authority in Giacomo Zabarella". *History of science* 45, pp. 404–427.

———. 2008. *Reenacting Galileo's experiments: Rediscovering the techniques of seventeenth-century science*. Lewiston, Queenston, Lampeter: The Edwin Mellen Press.

———. 2011. *A history of Galileo's inclined plane experiment and its philosophical implications*. Lewiston, Queenston, Lampeter: The Edwin Mellen Press.

Panofsky, Erwin. 1954. *Galileo as a critic of the arts*. The Hague: Martinus Nijoff.

Perifano, Alfredo. 1997. *L' alchimie à la Cour de Côme Ier de Médicis: savoirs, culture et politique*. Paris: Honoré Champion Éditeur.

Piccolino, Marco, & Wade, Nicholas, J. 2014. *Galileo's visions: Piercing the spheres of the heavens by eye and mind*. Oxford: Oxford University Press.

Plutarch. 1704. *Plutarch's Morals*. Translated by several hands. Volume I. London: Thomas Braddyll.

Poincaré, Henri. 1908. "L' invention mathématique". *Revue générale des Sciences pures et appliquées* 19, pp. 521–526.

Poppi, Antonino. 1992. *Cremonini e Galilei inquisiti a Padova nel 1604. Nuovi documenti d'archivio*. Padua: Antenore.

Prodi, Paolo. 1992. *Il sacramento del potere: il giuramento politico nella storia costituzionale dell'Occidente*. Bologna: Il Mulino.

Purnell, F. Jr. 1972. "Jacopo Mazzoni and Galileo". *Physis* 14, pp. 273–294.

Ragnisco, Pietro. 1894. "Da Giacomo Zabarella a Claudio Berigardo". *Atti del Reale Istituto Veneto di Scienze, Lettere ed Arti*, s. VII, 5, pp. 474–518.

Ramelli, Ilaria. 2005. *Corpus Hermeticum*. Edizione e commento di A. D. Nock e A.-J. Festigière. Edizione dei testi ermetici copti e commento di I. Ramelli. Milan: Bompiani.

Redondi, Pietro. 1994. "Dietro l'immagine. Rappresentazioni di Galileo nella cultura positivistica". *Nuncius* 9, pp. 65–116.

Reynolds, Anne. 1982. "Galileo Galilei's poem 'Against wearing the toga'". *Italica* 59, pp. 330–341.

Rhode, William S. 1971. "Observations of the vibration of the basilar membrane in squirrel monkeys using the Mössbauer technique". *The Journal of the Acoustical Society of America* 49, pp. 1218–1231.

Roda, Aldo. 2011. "A history of bioluminescence and chemiluminescence from ancient times to the present". In: Aldo Roda [Ed.]. *Chemiluminescence and Bioluminescence: Past, Present and Future*. Cambridge: The Royal Society of Chemistry Publishing, pp. 3–50.

Rösche, Johannes. 2008. *Robert Fludd: Der Versuch einer hermetischen Alternative zur neuzeitlichen Naturwissenschaft*. Göttingen: Vandenhoeck & Ruprecht.

Rossen-Knill, D. F., and Henry, R. 1997. "The pragmatics of verbal parody". *Journal of Pragmatics* 27, pp. 719–752.

Rothman, Aviva. 2009. "Forms of persuasion: Kepler, Galileo, and the dissemination of Copernicanism". *Journal for the History of Astronomy* 40, pp. 403–419.

Sawyer, R. Keith. 1992. "Improvisational creativity: An analysis of jazz performance". *Creativity Research Journal* 5, pp. 253–263.

———. 1996. "The semiotics of improvisation: The pragmatics of musical and verbal performance". *Semiotica* 108, pp. 269–306.

———. 2000. "Improvisation and the creative process: Dewey, Collingwood, and the aesthetics of spontaneity". *The Journal of Aesthetics and Art Criticism* 58, pp. 149–161.

Schmitt, Charles. 1969. "Experience and experiment: A comparison of Zabarella's view with Galileo's in *De motu*". *Studies in the Renaissance* 16, pp. 80–138.

Scribano, Emanuela. 2006. "Aristotele contro Platone in Fausto Sozzini". *Rinascimento* 46, pp. 73–92.

Scribano, Emanuela, & Priarolo, Mariangela. 2006. [Eds.] *Fausto Sozzini e la filosofia in Europa. Atti del convegno, Siena 25–27Novembre 2004*. Siena: Accademia senese degli Intronati (Il Leccio).

Sendivogius, Michael. 1628. *Novum lumen chymicum, e naturae fonte et manuali experientia depromptum, in duas partes divisum*. Geneva: Typis et sumptibus Ioan. De Tournes.

Serassi, Pierantonio. 1790. *La vita di Jacopo Mazzoni, patrizio Cesenate*. Rome: Stamperia Pagliarini.

Shea, William R. 1972. *Galileo's intellectual revolution*. New York: Science History Publications.

Sozzini, Fausto. 1610. *De statu primi hominis ante lapsum disputatio*. Cracow: Typis Sternacinianis.

Spini, Giorgio. 1983. *Ricerca dei libertini. La teoria dell'impostura delle religioni nel Seicento italiano*. Florence: Nuova Italia, 1983 (1st ed. 1950).

———. 1996. *Galileo, Campanella e il 'divinus poeta'*. Bologna: Il Mulino.

Strebel, Christoph. 2006. "Martinus Horky und das Fernrohr Galileis". *Sudhoffs Archiv*, Band 90, Heft I, pp. 11–28.

Targioni Tozzetti, Giovanni. 1780. *Notizie degli aggrandimenti delle scienze fisiche in Toscana nel corso di anni LX del secolo XVII*. 4 vols. Florence: Giuseppe Bouchard.

Trimegisto, Mercurio. 1549. *Il Pimandro di Mercurio Trimegisto, tradotto da Tommaso Benci in lingua Fiorentina*. Florence: Lorenzo Torrentino.

Truesdell, Clifford. 1976. "The scholar: A species threatened by professions". *Critical Inquiry* 2, pp. 631–648.

Utsumi, Akira. 2000. " Verbal irony as implicit display of ironic environment: Distinguishing ironic utterances from non irony". *Journal of Pragmatics* 32, pp. 1777–1806.

Vaccalluzzo, Nunzio. 1896. *Galileo letterato e poeta. Appendice, le rime inedite di Vincenzo Galilei*. Catania: Niccolò Giannotta.

———. 1910. *Galileo Galilei nella poesia del suo secolo. Raccolta di poesie edite e inedite scritte da' contemporanei in lode di Galileo, pubblicate in occasione del 3° centenario delle sue scoperte celesti*. Milan/ Palermo/ Naples: Remo Sandron.

Valla, Lorenzo. 1543. *Opera*. Basel: Apud Henrichum Petrum.

Vanini, Giulio Cesare. 1616. *De admirandis naturae reginae deaeque mortalium arcanis*. Paris: Apud Adrianum Perier.

Verene, Donald Phillip. 1981. *Vico's science of imagination*. Ithaca and London: Cornell University Press.

Vickers, Brian. 1983. "Epideictic rhetoric in Galileo's *Dialogo*". *Annali dell'Istituto e Museo di storia della scienza di Firenze* 8, pp. 69–102.

Vico, Giambattista. 1744. *Principi di scienza nuova*. 2 vols. Naples: Nella stamperia Muziana.

———. 1948. *The New Science of Giambattista Vico*. Translated from the third edition (1744) by T. G. Bergin and M. H. Fisch. Ithaca, New York: Cornell University Press.

Vocabolario della Crusca. 1623. *Vocabolario degli Accademici della Crusca, in questa seconda impressione da' medesimi riveduto, e ampliato, con aggiunta di molte voci degli autor del buon secolo, e buona quantità di quelle d'uso*. Venice: Iacopo Sarzina.

Vry, Silke. 2011. *Paul Klee for children*. Munich, London, New York: Prestel.

Westman, Robert S. 2011. *The Copernican question: Prognostication, skepticism, and celestial order*. Berkeley and Los Angeles: University of California Press.

Wilhelm, Richard, & Jung, Carl Gustav. 1931. *The secret of the golden flower. A Chinese book of life*. Translated and explained by Richard Wilhelm with a European commentary by C. G. Jung. London: Kegan Paul, Trench, Trubner & Co.

Wlassics, Tibor. 1974. *Galilei critico letterario*. Ravenna: Longo Editore.

Wootton, David. 2010. *Galileo: watcher of the skies*. New Haven: Yale University Press.

———. 2015. *The invention of science: A new history of the scientific revolution*. New York: Harper.

Yates, Francis. 2002. *Giordano Bruno and the Hermetic tradition*. London: Routledge (1st ed., Chicago: University of Chicago Press, 1964.)

Zabarella, Giacomo. 1608. *Opera logica*. Frankfurt: Sumptibus Lazari Zetzneri.

Zanella, Giacomo. 1868. *Versi*. Florence: G. Barbèra.

———. 1990. *Saggi critici*. 2 vols. Edited by Armando Balduino. Vicenza: Neri Pozza Editore.

INDEX

D

E

F

K

L

M

S

T

U

V

W

Y

Z

HISTORY AND PHILOSOPHY OF SCIENCE

Heresy, Crossroads, and Intersections

Paolo Palmieri, *General Editor*

This series invites book proposals that include innovative strategies for pursuing history and philosophy of science. Especially welcome are scholarly works using non-analytic philosophical perspectives to successfully bring to bear on our understanding of how scientific practices are related to the humanities and the social sciences. The series also welcomes exploration of the sciences in relation to gender, culture, society, and the intellectual and social contexts that illuminate the places, the structures of origination, and the patterns of development over generations. Approaches may include focused analyses of thinkers from unorthodox perspectives that can shed new light on the history and philosophy of science, such as Montaigne, Bruno, Galileo, Newton, Pascal, Emerson, Thoreau, Nietzsche, Jung, Freud. Proposals aimed at probing the philosophical intersections between the sciences and other societal practices that can be configured as heretic are also encouraged. These might include the emergence of the psychoanalytic movements in the twentieth century, how the fine arts have impinged on the historical processes that gave rise to the sciences over the last few centuries, how in turn the intellectual frameworks inaugurated by the sciences have been imported into the avant-garde movements that paralleled the advent of industrialized societies, and finally how contemporary scientific domains of knowledge reverberate in 'deviant' social and artistic practices.

For additional information about this series or for the submission of manuscripts, please contact:

Peter Lang Publishing
Acquisitions Department
29 Broadway, 18th floor
New York, NY 10006

To order other books in this series, please contact our Customer Service Department:

800-770-LANG (within the U.S.)
212-647-7706 (outside the U.S.)
212-647-7707 FAX

Or browse online by series at:

www.peterlang.com

www.ingramcontent.com/pod-product-compliance
Ingram Content Group UK Ltd.
Pitfield, Milton Keynes, MK11 3LW, UK
UKHW020214290926
15470UKWH00001B/189

9 781433 131400